PRAISE FOR
Consent of the Networked

"Nuanced and balanced.... *Consent of the Networked* is an excellent survey of the Internet's major fault lines."—*Wall Street Journal*

"An important new book.... [MacKinnon] argues that neither political action nor competitive pressure spawned by the free market will protect our rights, finally making a strong case for a third way—a nongovernmental watchdog with sufficient clout to preserve freedom on the Internet."
—*Boston Globe*

"[A] sharp, sobering rebuttal [to] heady rhetoric, questioning and complicating our understandings of what it means to be free online.... it should be required reading for anyone who cares about the future of the web—that is, for all of us."—*Mother Jones*

"An absolutely indispensable account of the way that technology both serves freedom and removes it.... MacKinnon does a fantastic job of tying her theory and analysis to real-world stories."—Cory Doctorow, *Boing Boing*

"MacKinnon offers a persuasive history of recent global protest movements, and her book serves as a primer on the role that Internet technology and by extension digital networks have played in those efforts."—The New Republic.com

"Facebook is now a semi-public space in which political and other potentially controversial views are expressed. Amazon is well on its way to becoming a dominant publisher. Google has the power to render any website effectively invisible. Given that freedom of speech is important for democracy, that means that these giant companies are now effectively part of our political system. Which is why we can expect *Consent of the Networked* to find its way on to reading lists in political science."—*The Guardian* (UK)

"Count me among those who are thoroughly convinced by MacKinnon's reporting and arguments . . . the people who should most read this book aren't the already aware, but folks—especially policy-makers—who see all the shiny devices and trendy social media and foolishly assume that the Internet will ultimately prevail. It might, but only if we understand what a lucky and unusual accident the Internet really is, and that to keep it open and free, we have to fight for it."—Micah L. Sifry, techPresident.com

"MacKinnon has rightly stressed the need for Internet users the world over to raise questions about the way technology is used in order to ensure that their rights and freedoms are protected."—*Truthout*

"MacKinnon manages to be critical without barnstorming over the details and without losing a sense of the promise of these technologies. She also is one of the first I have seen to suggest actionable solutions. . . . The book is well written, with a lushness of historical background and a breadth of learning that similar books sometimes lack."—*Ars Technica*

"Insightful and moving . . . Ms. MacKinnon's stories of the effort occurring worldwide as people harness the Internet, often with a political, socioeconomic or religious motivation, are discerning, harrowing and empowering. . . . Packed with thorough and impeccable research and persuasive, eye-opening anecdotes from around the world, *Consent of the Networked* should spearhead a robust debate."—*Washington Times*

"Ms. MacKinnon provides expert reporting and analysis of Internet censorship and acts against individuals by authoritarian regimes around the world. . . . [If] you are a Google user and don't understand their recently updated privacy policy, and are tired of trying to puzzle it out, then *Consent of the Networked* is for you."—*New York Journal of Books*

"Internet policy-making is fraught with contradiction, corruption, and colonialism. In *Consent of the Networked*, Rebecca MacKinnon has produced an incredibly well-researched account of these dilemmas, which is as deep as it is vast."—*Foreign Policy in Focus*

"A vitally important analysis of Internet manipulation that should be read by anyone relying on the web for work or pleasure."—*Booklist*

"An incisive overview of the global struggle for Internet freedom."—*Kirkus Reviews*

"[A] fascinating and provocative book."—*Publishers Weekly*

"[MacKinnon] makes a persuasive and important case that ... the effects of today's new communications systems, for human liberation or for oppression, will depend not on the technologies themselves but rather on the resolve of citizens to shape the way in which they are used."—James Fallows, National Correspondent, *The Atlantic*

"*Consent of the Networked* is a must-read for anyone interested in freedom of personal and political expression in the 21st century. It's accessible, engaging, and periodically hair-raising."—Anne-Marie Slaughter, Bert G. Kerstetter '66 University Professor of Politics and International Affairs, Princeton University

"The Internet poses the most complex challenges and opportunities for human rights to have emerged over the last decade. Rebecca MacKinnon's book is a clear-eyed guide through that complexity."—Mary Robinson, Former UN High Commissioner for Human Rights, and President of Ireland

"Rebecca MacKinnon has written a wonderfully lively and illuminating account of the issues we face in this contentious area [of cyber power and internet governance]. It is well worth reading."—Joseph S. Nye, Jr., University Distinguished Service Professor, Harvard University, and author of *The Future of Power*

"*Consent of the Networked* should be required reading for all of those involved in building our networked future as well as those who live in it." —Joi Ito, Director, MIT Media Lab

"A growing number of people throughout the world are counting on the Internet to move their countries in a more democratic direction. *Consent of the Networked* describes what's happening, successes and failures, what's next, and what needs to be done. It's the real deal."—Craig Newmark, founder of Craigslist

CONSENT
OF THE
NETWORKED

CONSENT

OF THE

NETWORKED

The Worldwide Struggle
for Internet Freedom

REBECCA MacKINNON

BASIC BOOKS
A Member of the Perseus Books Group
New York

Designed by Brent Wilcox

The Library of Congress has cataloged the hardcover as follows:
MacKinnon, Rebecca.
    Consent of the networked : the worldwide struggle for Internet freedom /
by Rebecca MacKinnon.
        p.   cm.
    Includes bibliographical references and index.
    ISBN 978-0-465-02442-1 (hbk. : alk. paper) —
ISBN 978-0-465-02929-7 (e-book)
    1. Internet—Political aspects.   2. Internet—Social aspects.
3. Internet—Censorship.   4. Freedom of information.   5. World
politics—21st century.   I. Title.   II. Title: world-wide struggle for
Internet freedom.
    HM851.M3327 2012
    302.23'1—dc23
                                                              2011036423

ISBN 978-0-465-06375-8 (paperback)

10 9 8 7 6 5 4 3 2 1

For Mom

Janice R. MacKinnon
1943–1999

CONTENTS

FOREWORD TO
THE PAPERBACK EDITION

This book was completed in August 2011. It went on sale five months later, in early 2012. The paperback edition will be released to readers in late spring 2013. Much has happened over those twenty months, to put it mildly.

Hardly a day goes by without at least one significant news headline related to issues of online free expression, electronic privacy, censorship, and digital surveillance in some part of the world. *Consent of the Networked* offers a broader conceptual framework for making sense of these developments: how they relate to one another, what they mean in the context of our own lives, and what each and every one of us can do to defend and preserve our rights in the Internet age.

The new afterword discusses some of the major events of 2012. Many readers have remarked that the book's arguments and analysis have not just been reflected in the events of late 2011 and all of 2012, but have indeed been reinforced by what has happened since the book first came out. As history continues to accelerate, I am hopeful that *Consent of the Networked* will remain useful to any reader concerned about the future of freedom in the Internet age.

Rebecca MacKinnon
January 2013
Washington, DC

PREFACE

One night in Beijing in 1998, I went to dinner with a few Chinese friends—all artsy intellectuals, journalists, and hip liberal types. As we dug into a mound of spicy Szechuan chicken and washed it down with lukewarm local beer, I described a book I had just finished reading: *The File* by Oxford historian Timothy Garton Ash. It tells the story of how after the Berlin Wall came down and the Iron Curtain crumbled, East Germans suddenly had access to the files the Stasi had been keeping on them. People found out who had been ratting on whom—in some cases neighbors and coworkers, but also lovers, spouses, and even sometimes their own children.

After I finished talking, one friend put down his chopsticks, looked around the table, and proclaimed, "Someday the same thing will happen in China. Then I'll know who my real friends are." The table fell silent.

If and when China does change, that day of reckoning may never come. In the age of broadband and smart phones, state security agents no longer depend so heavily on human informants. Not only are software and hardware more efficient and omniscient than humans, but conveniently they also lack consciences.

As a Beijing-based journalist working for CNN from 1992 to 2001, I found the Internet's arrival to be tremendously exciting. Most Western journalists and diplomats in China at the time assumed that the Communist Party's grip on power could never survive this new globally networked technology. A decade later, after leaving TV journalism and becoming an independent writer, researcher, and activist, I grew to realize that we were naive. Though the Internet has transformed Chinese society in many ways, the regime has also succeeded in adopting technology to its advantage in ways I had not imagined—ways I have spent much time over the past several years trying to understand.

After I left China in 2001 and moved to Japan as CNN's Tokyo bureau chief, my fascination with the Internet's effect on global politics broadened. On a trip to South Korea, I reported on how Roh Moohyun won the presidency in December 2002 by a narrow margin thanks to eleventh-hour online and mobile activism by readers of OhMyNews, one of the world's first online citizen journalism ventures. In January 2003 a friend introduced me to "Where is Raed?" the blog of an Iraqi man in Baghdad writing under the pseudonym Salam Pax. As the United States and its allies prepared to invade, he ranted cynically against Saddam Hussein's regime, the Bush administration, and almost everyone and everything else. As the web publishing magnate Nick Denton aptly put it, he was the "Anne Frank of the war . . . and its Elvis." As I started following blogs written by other less famous but no less eloquent people all over the world—people who were not professional journalists but who were witnesses or parties to events that no mainstream Western news media had reported—it was clear that the Internet-driven citizen media revolution had implications not only for the future of journalism but also for geopolitics.

In January 2004 I took what was supposed to have been five months' leave from CNN's Tokyo bureau to spend a semester at the Shorenstein Center on the Press, Politics, and Public Policy at Harvard's Kennedy School of Government, where I made it my full-time job to learn about the new world of citizen-driven online media. I started blogging. A couple of months into my leave, I decided to stay at Harvard and not to return to CNN. I moved to a new think tank called the Berkman Center for Internet and Society, the epicenter of new thinking and experimentation on citizen media and networked politics.

To keep my Chinese from getting too rusty, I began to follow Chinese blogs. Reading bloggers' observations, musings, and conversations helped me to keep up with events, cultural trends, and controversies in

China in a way that would have been impossible without the Internet. As it so happened, another colleague, Ethan Zuckerman, closely followed blogs from Africa and the Middle East. Ethan and I started to think about how we might support and amplify the work of bloggers who write about events in their countries, sharing insights and information not reported in the mainstream media. At the end of 2004 we organized a brainstorming meeting of bloggers from all over the world and called it Global Voices Online.

The discussions that day inspired our group to build an online community to help promote and support the work of civic-minded bloggers around the world. A conference blog evolved into a complex website. Today Global Voices is a full-blown nonprofit organization. Bloggers living all over the world work to curate, explain, summarize, and translate postings related to current events on blogs and other social media like Facebook, Twitter, and YouTube. Another team gives small grants to citizen media projects in the developing world. Yet another team translates Global Voices content back and forth across roughly two dozen languages. Last but not least, our activist team runs a network devoted to raising awareness about threats to online free expression and assembly, helping people navigate censorship and protect themselves against surveillance. As protests erupted in Tunisia in late 2010 and demonstrations spread around the Middle East and North Africa in early 2011, Global Voices contributors worked around the clock to spread information about what was happening in multiple languages, on our own site as well as Twitter, Facebook, and other social media platforms.

As Global Voices grew, I continued to follow the evolution of Chinese blogs and social media, returning regularly to China to attend blogger gatherings and getting to know some of China's boldest online activists. I also began to research China's Internet censorship system. Other researchers had begun to study China's "great firewall"—the system by which a large number of overseas websites are blocked from view on the Chinese Internet. I soon realized, however, that website blocking was just one part of the Chinese government's strategy to control the online activities of Chinese citizens. A core component of that strategy involves censorship and surveillance carried out not by government agents or "Internet police" but by the private sector. After Yahoo, Microsoft, Cisco, and Google came under fire at a 2006 congressional hearing for their role in Chinese censorship and surveillance, I began to be asked regularly to write and speak about the problem of corporate collaboration in Chinese censorship and surveillance. I joined a group of Internet companies, human rights organizations, socially responsible investors, and academics who together in 2008 launched the Global Network Initiative, an organization that has established standards for free expression and privacy for the Internet and telecommunications sector, to which it seeks to hold companies accountable.

Five years ago when I first considered writing a book, I was inclined to focus on China as "exhibit A" for how an authoritarian regime can not only survive, but also thrive in the Internet age with the help of domestic and multinational corporations. My experience with Global Voices and as an advocate for online free speech on an international scale, however, has also taught me that the root causes are broader. The technologies and policies that make surveillance and censorship possible in China and many other countries are closely connected to policy, business, and technical decisions being made by governments and companies in the democratic West. Sometimes those decisions are made by people who understand the implications of their actions but simply have other priorities. Others have good intentions but are ill-informed about the dynamics of power, control, and freedom across a global Internet.

In mid-August 2011, as I completed the final edits on this book, riots broke out in Great Britain. Debates raged in the United Kingdom over Prime Minister David Cameron's controversial remarks about the need for expanded government power to monitor and restrict the British public's access to mobile services as well as to social networks. In the United States, San Franciscans were up in arms after the local subway system, Bay Area Rapid Transit (BART), shut down wireless service at several stations to prevent a planned protest against a shooting by BART police of an allegedly knife-wielding man. China's state-run

Xinhua News Agency could not resist the opportunity to gloat: "We may wonder why Western leaders, on the one hand, tend to indiscriminately accuse other nations of monitoring, but on the other take for granted their steps to monitor and control the Internet," said the unsigned commentary. "For the benefit of the general public, proper Web monitoring is legitimate and necessary."

I am concerned about what will happen to the Internet—and more broadly to the future of freedom in the Internet age—if the world's democracies develop a habit of tackling problems in a short-sighted, knee-jerk manner, without considering the long-term domestic and global consequences. This book is thus a distillation of what I have learned over the past decade about the global struggle for Internet freedom and an attempt to apply those learnings to our common future. It is an effort to crystallize what worries me most—and to galvanize others to defend individual rights that are under threat, on a scale and in ways that are only beginning to become apparent. I worry that the future of freedom and democracy in the Internet age will not be as bright as many assume. Yet I am also hopeful: I have had the privilege to work with innovative and often very brave people from all over the world who are using technology to speak truth to power and to give voice to their communities and fellow citizens.

Many of these people are fighting uphill battles just to be able to keep using technology to express themselves and organize peacefully without censorship or fear of reprisal. The obstacles to their success are created not only by authoritarian governments, but also by Western companies and democratically elected politicians who do not understand the global impact of their actions—or, more ominously, do not care. This book is my effort to raise the level of public awareness about the many inconvenient truths of the Internet age. I have sought to explain what I believe all citizens everywhere need to know about the global struggle for Internet freedom. The outcome of this struggle will affect each and every one of us. It is a struggle that all of us have the power and ability to influence—even in small ways—if we understand the complex forces at work and how we might shape them.

Entire books have been written about many of the issues raised in each chapter. I have listed some of them in the endnotes for readers who want to go deeper. My aim here is to tie together the broader set of overlapping, complex issues in a way that makes sense to an informed reader who does not have special Internet-related expertise—beyond simply being an Internet and cell phone user. For people who are experts on some of these issues, I have tried to provide a fresh conceptual framework and geopolitical context, which I hope will be useful to experts and nonexperts alike who are concerned about the future of freedom in the Internet age.

It is not possible to document in one concise book all the violations of Internet freedoms and rights happening everywhere in the world. If your rights to digital free expression and assembly are under attack but your country is not mentioned in this book, please understand that the omission does not imply a lack of concern for the violations you and your compatriots are enduring. Several organizations, including the OpenNet Initiative and Freedom House, produce regular reports systematically documenting attacks on Internet freedom, country by country. Nor is it possible in one book to catalog every violation of free speech and privacy by every company everywhere. I have selected specific examples that best support the overall argument. Please see the book's companion website at www.consentofthenetworked.com for further sources of information and links to organizations that document ongoing cases.

I am not able to list here all of the people who made this book possible—or who taught me what I needed to know to write it. Beyond the obvious reasons to thank my parents, I also owe most of the credit for my fluency in Chinese, lifelong connection to China, and broader fascination with global politics to Stephen R. MacKinnon, professor of modern Chinese history, and the late Janice R. MacKinnon. In 1979 my parents took my brother, Cyrus, and me to live in Beijing for two years, enrolling us in a Chinese primary school, so they could conduct research on a book. None of us realized at the time what a gift they had given their children.

More than two decades later when I was working as a foreign correspondent in Tokyo, I met Joi Ito, who introduced me to blogs, catalyzing a chain of discoveries and friendships that ultimately propelled me away from conventional journalism and down the winding path I have taken. Adam Greenfield jolted me out of my TV reporter's mindset and challenged me to think in new dimensions. My dear friend and colleague Ethan Zuckerman has been a rock since we started collaborating in 2004, providing critical support and substantive advice as this project evolved. Harvard's Berkman Center for Internet and Society is a truly unique intellectual launching pad and support system from which I continue to benefit in delightfully random ways. In 2007 and 2008, Ying Chan at Hong Kong University's Journalism and Media Studies Centre took me in and supported my controversial research and unorthodox projects.

A generous fellowship from the Open Society Foundations in 2009 supported research and travel that enabled me to figure out what this book really needed to be about. Without subsequent fellowships from Princeton's Center for Information Technology Policy and the New America Foundation in Washington, DC, I could not have afforded the time and head space that completion of this project demanded. At various points along the way, friends and colleagues gave me ideas that I have incorporated. Notably, "digital bonapartism" came from the great brain of Yale historian Timothy Snyder. Joshua Cooper Ramo and Kathy Robbins offered game-changing reactions to early versions of the book proposal. Family and friends lent guest bedrooms, proffered food and drink, and provided moral and emotional support throughout my nomadic research and writing phase. Henrik Bork and Wang Kuangyi were especially generous with their Beijing apartment. Most profoundly, the Global Voices community schooled me daily, giving me a faith in humanity that I might otherwise have been hard-pressed to retain.

It has been a delight and an honor to be represented by John Brockman, visionary curator of ideas and convener of people working at the intersection of science, technology, and society. He grasped what I was trying to do before I completely understood it myself, challenged me to focus and refine my argument, and made it possible for Max Brockman to deploy his literary matchmaking talents. I am eternally thankful to TJ Kelleher at Basic Books for believing in me, and for guiding me through the muddy waters of rough drafts and uncertainty as a first-time book author. Special gratitude goes to Ethan Zuckerman, Maya Alexandri, Ian Johnson, Tom Glaisyer, Ivan Sigal, David Sasaki, Luisetta Mudie, Mary Kay Magistad, Anna Husarska, and my father, Stephen MacKinnon, who volunteered precious time to read drafts and comment. Finally, I have no idea how I would have held on to my sanity in the process of intense writing, editing, and rewriting without the support—editorial and emotional—of my loving partner and intellectual muse, Bennett Freeman.

August 2011
Washington, DC

INTRODUCTION

After the Revolution

On March 5, 2011, protesters stormed the Egyptian state security head-quarters. In real time on Twitter, activists shared their discoveries with the world as they moved through a building that had until recently been one of the Mubarak regime's largest torture facilities. Videos and photos uploaded to YouTube, Flickr, and Facebook showed a flurry of young men (and a few women) opening doors and cabinets, sifting through piles of shredded paper, pulling out stacks of files, and examining pieces of equipment. Some were implements of torture.

"Entered the small compound where I was locked," tweeted Hossam el-Hamalawy, a thirty-three-year-old journalist and activist who had been detained and tortured several times since they first picked him up as a student activist thirteen years ago. Returning home a few hours later, he told his followers, "I've been crying hysterically today."

Some activists found their own files. They were full of wiretap transcripts, reams of printouts of intercepted e-mails and mobile messages. All kinds of records had been kept about them: lurid details of divorces and personal relationships; all their past job applications; foreign organizations they had communicated with; international meetings they had attended. Clearly the Egyptian government had sophisticated surveillance technology at its disposal. It still does.

Wael Ghonim, the young Google executive and a hero of the Egyptian revolution for his role in creating and running the Facebook protest group that played a key role in getting the first wave of protesters into

Cairo's Tahrir Square, famously told CNN, "If you want to liberate a society just give them the Internet." The Internet certainly did play a powerful role in bringing down a dictator—the Internet in the hands of a committed community of activists who spent the better part of a decade building a movement. It is less clear how helpful the Internet will be when it comes to protecting the Egyptian people's rights in the post-Mubarak era and in building a new democracy.

If the events of 2011 taught the world anything, it is that although the Internet empowers dissent and activism, it is not an instant freedom tonic that, when applied in sufficient quantities, automatically results in freedom. It is time to stop debating *whether* the Internet is an effective tool for political expression, and to move on to the much more urgent question of *how* digital technology can be structured, governed, and used to maximize the good it can do in the world, and minimize the evil.

Even if Egypt does manage to establish a stable democracy, the Egyptian people face a problem that none of the world's democracies have yet solved: How do we make sure that people with power over our digital lives will not abuse that power?

In mature democracies, laws and regulations governing digital platforms and networks arguably represent "consent of the governed"—to the extent that political traditions and institutions reflect the will of the people as a whole, as distinct from representing mainly special interest groups, corporate lobbies, and the loudest and most extreme elements of both political parties. Of course, the extent to which democratic politics in the United States, United Kingdom, and other democracies around the world actually succeed in accomplishing the former rather than the latter is a matter of fierce debate.

Americans across the political spectrum continue to have real and legitimate concerns that the system—the political process, media organizations, and information networks—is vulnerable to capture, manipulation, and abuse. There is good reason to worry about the way in which our digital lives are being shaped by regulators, politicians serving powerful constituencies, and companies seeking to maximize profit. The potential for manipulation and abuse of the digital networks and platforms that citizens have come to depend upon is one of the more insidious threats to democracy in the Internet age. If citizens of established democracies cannot prevent such manipulation, the prospects for aspiring and fragile democracies in Tunisia and Egypt, let alone hopes for the future in places like Iran and China, look much less bright.

It is a reality of human nature that those with power, however benign or even noble their intentions, will do what they can to keep it. The American revolutionaries certainly understood this truth. As James Madison, one of the framers of the Constitution, observed near the end of his life in an 1829 speech before the Virginia Constitutional Convention, "The essence of Government is power; and power, lodged as it must be in human hands, will ever be liable to abuse." This was their fundamental assumption as they formed new institutions of the world's first democratic republic.

In the physical world, mechanisms of democratic politics and constitutional law have worked—not perfectly, but still far better than any alternatives—to protect citizens' rights. But these mechanisms are no longer adequate for people whose physical lives now depend on what they can or cannot do (and what others can do to them) in the new digital spaces where sovereignty and power are ill-defined and highly contested. The reality is that the corporations and governments that build, operate, and govern cyberspace are not being held sufficiently accountable for their exercise of power over the lives and identities of people who use digital networks. They are sovereigns operating without the consent of the networked.

This absence of consent takes place on several levels. Governments are exercising power over people outside their jurisdictions through global Internet companies. When citizens depend on platforms like Google, Twitter, and Facebook, whose digital inhabitants include people from nearly everywhere on earth, legislators and regulators in the world's largest markets make decisions that ultimately shape global technical standards and business norms. Thus governments are exerting power over the freedoms and rights of people who did not vote for them, who do not live under their jurisdiction, and who have no meaningful way of holding them accountable.

"Governance" functions, once carried out almost entirely by nation-states, are now shared increasingly by private networks and platforms. The lives of people around the world—from people living in democracies, such as the United States, to those living in authoritarian regimes, such as China—are increasingly shaped by programmers, engineers, and corporate executives for whom nobody ever voted and who are not accountable to the public interest in any way. When we sign up for web services, social networking platforms, broadband service, or mobile wireless networks, and we click "agree" to the terms of service, we give them false and uninformed consent to operate as they like.

Today the power of Internet platforms and services to shape people's lives is greater than ever and will only continue to grow. Though some Internet companies are helping to empower people in innovative and exhilarating ways, many are also helping authoritarian dictators adapt and survive in the Internet age. In China, Western companies and financiers have supported and helped to legitimize a political innovation that I call "networked authoritarianism," in which corporate networks are turned into opaque and subtle but invasive extensions of government power. Leaders of stunted crypto-democracies like Russia have adopted what I call "digital bonapartism," using populist rhetoric, combined with control over private enterprise and the legal system, to marginalize the opposition and manipulate public opinion much more subtly than in the old days. Even theocratic dictatorships like Iran are using their control over networked technologies to serve their economies and improve governance while preventing opponents from successfully using the Internet to overthrow them.

Many corporate executives argue that human rights are neither their concern nor their responsibility: the main obligation of any business, they point out, is to maximize profit and investor returns. But what kind of world are they helping to create, and should that not concern them? In the first half of the twentieth century, corporations were forced to consider employees' safety and health. In the past fifty years the environmental movement has forced industry to share responsibility for pollution and more recently climate change—though they have not gone nearly far enough. So far, most Internet and telecommunications companies have failed to accept responsibility—beyond cyber-utopian platitudes—for the rights of their customers and users, even as companies in other, much older industries have long since begun to do so with their workers, shareholders, and broader stakeholders.

Building accountability into the fabric of cyberspace requires political innovation to match the rapid technical innovation of the forty-plus years since the Internet was invented. In the twenty-first century, many of the most acute political and geopolitical struggles will involve access to and control of information. Human freedom increasingly depends on who controls what we know and therefore how we understand our world. It depends on what information we are able to create and disseminate: what we can share; how we can share it; and with whom we can share it. It also depends on the extent to which we have any control—or any say in—how our own information is shared with other people, private companies, and governments.

This book is about the new realities of power, freedom, and control in the Internet age. People, governments, companies, and all kinds of groups are using the Internet to achieve all kinds of ends, including political ones. But we cannot understand how the Internet is used unless we first understand the ways in which the Internet *itself* has become a highly contested political space. Pitched battles are currently under way over not only who controls its future, but also over its very nature, which in turn will determine whom it most empowers in the long run—and who will be shut out.

Contrary to what some people may have hoped and believed, the Internet does not change human nature. We have begun to see how absolute power in cyberspace corrupts absolutely as it does in physical space. As with power in the physical world, power in the digital world must be constrained, balanced, and held accountable. The future of freedom in the Internet age depends on the choices and actions of everybody on the planet who creates, uses, and regulates technology. It depends on whether we assert our rights within the digital spaces we now inhabit—just as our forebears fought for their rights in the physical spaces once controlled entirely by sovereigns who claimed to have the divine right to rule as they pleased.

The first step toward engaging in this struggle is to understand how the dynamics of power and freedom have changed, now that our political lives have become highly dependent on digital services and platforms that are largely owned and operated by the private sector. Part One will explain how the Internet—driven by the private sector—has challenged the power and legitimacy of the nation-state, and has also given rise to a new digital "commons" that incubates much innovation as well as some of the world's most disruptive digital activism. Part Two describes a phenomenon that some call "Control 2.0": how opaque, unaccountable relationships with Internet and telecommunications companies enables authoritarian governments to control and manipulate citizens. Though China is the most advanced case, a range of other authoritarian regimes are also taking advantage of their power over private networks and platforms. Part Three examines how democracies are being corroded by increasingly opaque and unaccountable relationships between government and the companies that own and operate the digital networks and platforms upon which our democratic discourse increasingly depends. Part Four describes how companies act as the new sovereigns of cyberspace—and how most companies' failure to take responsibility for their power over citizens' political lives, and their lack of accountability in the exercise of that power, corrodes the Internet's democratic potential in often subtle and insidious ways. Finally, Part Five explores efforts by some governments, a few companies, and a growing number of concerned citizens to address the threats to freedom in cyberspace through new initiatives and movements.

The Internet is a human creation. Power struggles are an inevitable feature of human society. Democracy is about constraining power and holding it accountable. The Internet can be a powerful tool in the hands of citizens seeking to hold governments and corporations to account—but only if we keep the Internet itself open and free.

PART ONE

DISRUPTIONS

To no one will we sell, to no one will we refuse or delay right of justice.

—MAGNA CARTA, 1215

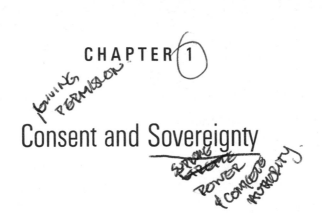

CHAPTER 1

Consent and Sovereignty

On Super Bowl Sunday, January 22, 1984, Apple ran one of the most famous TV advertisements of all time. It opened with a gray theater full of people with shaved heads, wearing gray jumpsuits, staring expressionlessly at a large screen. From the screen, an Orwellian "Big Brother" intoned, "We are one people, one whim, one resolve, one course. Our enemies shall talk themselves to death, and we shall bury them with their own confusion. We shall prevail." As he spoke, an athletic young blonde woman in a blinding-white tank top and bright orange running shorts ran into the theater and down the center aisle, carrying a sledgehammer. She threw it at the screen, and the screen exploded. An off-camera voice declared, "On January 24th, Apple Computer will introduce Macintosh. And you'll see why 1984 won't be like 1984."

Today, more than two decades later, the message remains tremendously powerful: innovative technology in the hands of brave people can free us all from tyranny. Apple updated the commercial for its January 2004 MacWorld Expo, adding an iPod and earbuds to the outfit of the sledgehammer-wielding athlete.

The following month, a Tunisian lawyer and human rights activist named Riadh Guerfali, known publicly before his country's 2011 revolution only by his pseudonym, Astrubal, uploaded a mash-up of the ad onto the video-sharing platform Dailymotion. He replaced the on-screen Big Brother with video of President Zine El Abidine Ben Ali. After the athlete sledgehammers the screen, the screen goes white and the video cuts to a Tunisian girl with her eyes shut. She opens her eyes, as if waking from a bad dream. The video ends.

Guerfali's video was part of a broader digital activism campaign that he and a group of Tunisian activists launched in 2002, before YouTube was invented and before Facebook and Twitter were even twinkles in their creators' eyes. Their strategy was to counter the constant stream of government propaganda with clever antigovernment "propaganda" of their own.

In response, the Tunisian government developed the Arab world's most sophisticated censorship regime. But censorship was not the only way in which the Tunisian people's digital rights were regularly and systematically violated. Digital surveillance in Tunisia was even more pervasive than in Egypt. Progovernment hackers attacked dissident websites with aggressive "denial of service attacks" and took them offline. Government-controlled companies that provide Internet service to offices and homes used "deep packet inspection" technologies to track and filter everything passing through their networks. Government-employed geeks hacked into activists' computers and stole information, intercepted and even altered people's e-mails, and took over activists' Facebook accounts by intercepting their passwords.

After the revolution, several former dissidents and government critics were brought into Tunisia's transitional cabinet—including the dissident blogger Slim Amamou, who had spent the final weeks of Ben Ali's reign in jail for his digital activism. But arguments quickly arose over exactly how free Tunisia's information networks ought to be. Less than a week after the new post–Ben Ali government had formed, Tunisian State Secretary Sami Zaoui announced that the government would continue blocking websites deemed to be "against decency, contain violent elements or incite to hate." In response to fierce backlash by the Tunisian free-speech movement, he retorted on Twitter, "Wrong! Even the countries that are most evolved when it comes to freedom block terrorist sites." Five months later, the government blocked a number of Facebook pages and groups, citing concerns over inflammatory and offensive speech. Amamou resigned in frustration. Guerfali commented philosophically, "Before things were simple: you had the good guys on one side, and the bad guys on the other. Today, things are more subtle."

George Orwell published *1984* at the dawn of the Cold War, as a warning about the totalitarian possibilities of a modern industrialized state that combines centralized power, utopian ideology, and electronic media. The struggle for freedom in the Internet age is shaping up to be very different from the ideological struggles of the twentieth century. Today's struggle is not a clear-cut contest of democracy versus dictatorship, communism versus capitalism, or one ideology over another. Human society has acquired a digital dimension with new, cross-cutting power relationships. The Internet is a politically contested space, featuring new and unstable power relationships among governments, citizens, and companies. Today's battles over freedom and control are raging simultaneously across democracies *and* dictatorships; across economic, ideological, and cultural lines.

Internet platforms and services, made commonplace by companies such as Apple, Google, Facebook, and Twitter, along with a range of mobile, networking, and telecommunications services, have empowered citizens. They have empowered us to challenge government, both our own as well as other governments whose actions affect us. But the Internet also empowers governments themselves—or at least the growing number whose police, military, and security forces understand how the Internet works and who have learned the value of employing computer-science graduates. *All* governments, from dictatorships to democracies, are learning quickly how to use technology to defend their interests.

Apple's dramatic 1984 Super Bowl ad notwithstanding, in reality the interests and loyalties of corporations are divided. On the one hand are the customers and users—also citizens of polities—whose trust is required for long-term business success, and who themselves hold a range of often-conflicting beliefs and values. On the other hand are governments, whose approval and regulatory support is critical if the corporations are to run profitable businesses or gain access to lucrative markets, and who are often important customers themselves. In an ideal world, the government would serve citizens' interests and ensure that their rights are protected. In the real world, we are not so naive as to assume this is the case, certainly not in authoritarian dictatorships and, depending on one's political viewpoint, not always in democracies either.

The problem is that our ability to organize and speak out is shaped—often quite subtly—by the Internet service providers, e-mail services, mobile devices, and social networking services. If our communications and access to information are manipulated in ways we are not aware of, and if these companies' relationships with government are opaque, our ability to understand how power is being exercised over us, and our ability to hold that power to account, will be eroded in a more subtle and insidious manner than Orwell ever imagined.

In the Internet age, the greatest long-term threat to a genuinely citizen-centric society—a world in which technology and government serve citizens instead of the other way around—looks less like Orwell's *1984,* and more like Aldous Huxley's *Brave New World*: a world in which our desire for security, entertainment, and material comfort is manipulated to the point that we all voluntarily and eagerly submit to subjugation. If we are to avoid this dystopian fate, political innovation will have to catch up with technological innovation.

CORPORATE SUPERPOWERS

In an April 2011 town hall meeting at Facebook's Palo Alto headquarters, President Barack Obama waxed enthusiastic about the political power of social networking:

> Historically, part of what makes for a healthy democracy, what is good politics, is when you've got citizens who are informed, who are engaged. And what Facebook allows us to do is make sure this isn't just a one-way conversation; makes sure that not only am I speaking to you but you're also speaking back and we're in a conversation, we're in a dialogue. So I love doing town hall meetings. This format and this company I think is an ideal means for us to be able to carry on this conversation.

For any citizen who cares about politics, this sounds very exciting. But when we step back and think about it more broadly, it is unclear what the president's growing reliance on Facebook for political campaigning and policy advocacy ultimately means for American democracy. Will the result be a net plus for improving American democracy and making government more accountable? Or will a codependency between politicians and social networking platforms result in ever more subtle manipulation of the public discourse?

Meanwhile, Facebook, like any large organization with global interests, has been working hard to win friends and influence people in Washington and other world capitals, to maximize its profit-making potential. In 2010 and 2011, the company beefed up its Washington-based policy team, not only to lobby against regulations that the company believes hamper its interests, but also to help politicians integrate Facebook as a key tool for election campaigns as well as for constituent communications. In May 2011, the company also announced that it would dispatch emissaries to major world capitals to "promote the uses of Facebook with policymakers and influencers in both electoral and governing bodies," and "monitor legislation and regulatory matters." After all, with 600 million users by mid-2011, the social network has more users than most countries have citizens. Facebook, however, was not the first company to create a quasi-diplomatic corps. Since 2005, Google has been hiring executives with government and diplomatic experience for positions described internally as "foreign minister" and "ambassador."

The new digital superpowers have begun to clash with conventional nation-states. A classic example was Google's clash with the Chinese government. In March 2010, Google stopped censoring its Chinese search engine, Google.cn, and moved it out of mainland China in response to aggressive and sustained attacks launched from Chinese computer servers on Google's Gmail service a few months before. The Chinese government denied knowledge of or connection with the attacks, denials that security experts and Western diplomats found difficult to believe given the attacks' military-grade sophistication. Plus Gmail also happened to be the e-mail service of choice for Chinese dissidents and activists, in addition to being

popular among businesspeople, computer programmers, academics, and students. An editorial in China's state-run *People's Daily* responded to Google's defiance of censorship and withdrawal by calling the company "a tool of the US to implement its Internet hegemony." If Google was not going to obey censorship orders and respect Chinese law, officials said, then good riddance. Yet in the end, Google was not fully banned from China. It retained its license to keep a business presence in China and continued some activities not related to search: Android mobile phone operating system development and support, advertising sales, plus research and development for future products.

The reason has to do with Google's own Chinese constituency: people who need access to at least some of Google's products and services to do their jobs and build their own innovative businesses. In the first three months of 2010, when Google's fate in China remained uncertain, people in Beijing and Shanghai laid flowers outside Google offices, as if to commemorate the death of a relative. Many people commented in chat rooms and on social networks, and even told reporters, that they did not know how they could do their jobs effectively or keep up on the latest research in science and technology around the world without access to Google. Heavy reliance on Google services among foreign businesspeople and investors in China, plus Google's appeal to a certain cohort of educated, professional, urban Chinese—and particularly among China's own technical and business community—was believed to be a big reason the Chinese government did not order a block on all Google services (including Gmail, Google Docs, Google Scholar, and Google Reader) that year.

Some influential Chinese businesspeople argued that an outright ban on Google was simply not in the interest of Chinese industry. Though most expressed this view in private, one prominent member of the business community, Edward Tian, founding CEO of China Netcom, China's first broadband company, spoke out in a public conference, asking, "When we make this sort of company such a big rival, are we not also rejecting these technologies?" Banishing one of the world's most innovative companies from China, he warned, could hurt Chinese companies' ability to innovate and compete in the global marketplace. As the Chinese blogger Michael Anti commented to me wryly at the time, "Google is much more popular in China than the USA." Almost no Chinese citizens consider themselves stakeholders or constituents of the United States. But Google has many Chinese constituents. They are, in effect, digital residents of Googledom: a global community of people who rely on certain Google services.

In late 2010, Google CEO Eric Schmidt and Jared Cohen (who had just left the State Department policy planning staff in the summer of 2010 to run Google's new policy think tank) published an article in *Foreign Affairs* outlining their geopolitical vision for a digitally networked world. "Democratic governments," they wrote, "have an obligation to join together while also respecting the power of the private and nonprofit sectors to bring about change." They warned against overregulation of Internet companies, lest their greatest value to citizens be stifled.

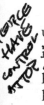

Google and Facebook are just two of the many companies whose products and services have created a new, globally networked public sphere that is largely shaped, built, owned, and operated by the private sector. Digital platforms, services, and devices now mediate human relationships of all kinds, including the relationship between citizens and government. Struggles to control and shape this sphere are intensifying around the world. These struggles will only escalate as the stakes continue to rise.

There is no question that the Internet would not be what it is today, and could not fulfill its potential as an empowering force for the disaffected, ignored, and oppressed, without a vibrant and innovative private business sector. Successful businesses need regulatory and legal environments that allow private citizens to form companies, borrow money or obtain investment, and protect their rights to property and inventions— both physical and virtual. It was no accident that Microsoft, Apple, Google, and Amazon were spawned by entrepreneurs in the United States and not in other countries with more constrained regulatory, entrepreneurial, and political environments.

Many people around the world are concerned, however, that Internet and telecommunications companies have gained far too much power

over citizens' lives, in ways that are insufficiently transparent or accountable to the public interest. In just one of many examples, German
and Korean citizens were particularly outraged by what they perceived
to be an unjustified invasion of their privacy by Google Street View, a
global service that enables people to zero in on a map to see in minute
detail what any given street or neighborhood looks like. In April 2011,
researchers exposed that Apple iPhones were logging and storing detailed information about users' movements, unbeknownst to most
iPhone users. (Apple later fixed what it described as a "bug" in the
phone's operating software.)

Companies argue that collecting a wide array of personal data is necessary to serve people better, in ways most people have shown that they
want. Critics argue that companies have gone far beyond what most citizens *actually* want—when they have a chance to understand what is really going on. In his book *The Filter Bubble*, Eli Pariser warns that search
engines and social networks manipulate what we find and who we in-
teract with on the Web in a way that maximizes our value to advertis-
ers but that is likely to minimize the chances that we will be exposed to
a sufficiently diverse range of news and views that we need as citizens to
make informed political and economic choices. In *The Googlization of
Everything*, Siva Vaidhyanathan warns that Google in particular represents a new ideology that he calls "techno-fundamentalism," which en-
courages a dangerously "blind faith in technology" on the part of people
who use Google services. Such faith, he argues, blinds us to what companies might be doing differently, how their internal decisions affect
our lives in ways we have never thought about, and whether our excitement with new technologies lulls us into accepting risks that we do not
see or understand.

The geopolitical power of corporations has been growing for decades
and is certainly not limited to Internet-related companies. As Harvard's
Joseph Nye points out in *The Future of Power*, when a corporate giant
like IBM derives two-thirds of its revenue from outside its home US
base, with only one-quarter of its workforce living in the United States,
the conventional power politics of nation-states is disrupted by the

emerging power of the private sector. Approximately half of the world's one hundred largest economic entities are now corporations. If Walmart were a country, it would be the world's twenty-fifth-largest economy in terms of GDP, ahead of countries such as Norway, Venezuela, and the United Arab Emirates.

Other kinds of transnational organizations are also challenging the power of nation-states. International institutions such as the United Nations, the World Bank, and the International Monetary Fund, set up at the end of the Second World War, use the nation-state as their organizing principle, while the world's power dynamics increasingly are driven by a much more diffuse set of actors. Newer organizations like the World Economic Forum, along with so-called multi-stakeholder corporate responsibility initiatives, create spaces where states, nongovernmental organizations (NGOs), and corporations can meet if not as equals, then at least as joint actors based on the premise that solutions to global challenges are beyond the legitimacy and capacity of any one sector to succeed.

It is now normal for the world's most powerful governments to consult with multinational corporations to shape a range of financial, trade, and foreign policy objectives. Not only are the world's most powerful democracies prodded and shamed by international NGOs like Human Rights Watch, Global Witness, Oxfam, and Greenpeace, but governments also increasingly find themselves needing to coordinate with— and even answer to—these and many other organizations in addressing a range of human rights and developmental issues. Companies and NGOs now show up in force to lobby their interests and causes at UN meetings on problems ranging from climate change to the new and thorny question of Internet governance.

Internet-related companies are even more powerful because not only do they create and sell products, but they also provide and shape the digital spaces upon which citizens increasingly depend. Governments of all kinds seek to control them precisely because of this power. Amid such dramatic changes in the power dynamic, it is important to remember the original purpose of democratic government and politics: to ensure that citizens' interests are served and that their rights are protected.

LEGITIMACY

The idea of the nation-state as the main organizing unit for politics and geopolitics, and the further notion that nation-state governance should be grounded in "consent of the governed," are both relatively new ideas that did not spread on a truly global scale until the late twentieth century. Modern concepts of sovereignty and legitimacy first germinated eight hundred years ago when a critical mass of English nobility decided that the unconstrained "divine right of kings" no longer served their interests, and realized that they had the economic and military power to do something about the situation. In 1215 at a field called Runnymede, the English nobility, fed up with arbitrary jailings, confiscations of property, and what they felt were other unreasonable violations of their rights, forced King John to sign a document called the Magna Carta. This "great charter" recognized that even kings must be bound by law and unleashed a powerful set of ideas about legal rights and political sovereignty.

The Magna Carta was the first attempt in the Western world since the ancient Greeks to enshrine the idea that a sovereign's legitimacy does not derive solely from divine right and brute force; even the king is subject to the law. Though the barons were hardly populists, let alone revolutionaries—their goal was to protect the interests of an elite ruling class—the Magna Carta nonetheless laid the groundwork for the concept of the modern nation-state, based on the rule of law over men.

It would take four more centuries for the sovereignty of the modern nation-state to be formalized in Europe by the 1648 Treaty of Westphalia. Political philosophers from Hobbes to Locke to Rousseau expanded and broadened the idea of "consent of the governed." In 1647 during the English civil war, a scrappy group of commoners calling themselves the Levellers issued a declaration of rights, including assertions that all citizens have the right to liberty of conscience and religion; that all laws must respect all persons equally, regardless of wealth, nobility, or lack thereof; and that Parliament must pass no law "evidently destructive of the safety or wellbeing of the people."

The Levellers debated their allies against King Charles I in Putney Church on the banks of the Thames: a physical commons for political discourse in which sovereignty and consent were freely debated in the most radical terms yet in English history. Even though they failed, their ideas eventually inspired the American Declaration of Independence and the Bill of Rights, and became the building blocks of the United Kingdom's modern parliamentary system. The Magna Carta is the antecedent and the Levellers are the ancestors of democracy movements and human rights struggles around the world. They are also the progenitors of modern struggles for civil liberties in today's democracies, as citizens never stop fighting to bring their societies closer to their founding ideals.

The Internet holds immense potential to help citizens improve democratic governance where it already exists, and has proven to be an effective tool in the hands of activists in exposing injustice and even overthrowing dictators. It offers a global platform for public discourse on matters of sovereignty and consent—yet those ideals are contested and even imperiled within the digital commons.

In his 2009 book, *Communication Power*, Internet scholar Manuel Castells recounts the numerous victories of popular movements and Internet-powered grassroots campaigns around the world over the past decade, describing how "insurgent communities" have managed to "reprogram" national politics in many countries. Yet he ends with a warning: digitally empowered citizens may have won important victories, but these victories are not necessarily permanent "because the powerholders in the network society" will do everything possible to "enclose free communication in commercialized and policed networks."

We would do well to heed Castells's warning. As the Internet grows ever more intertwined with our lives, citizens' dependence on it for achieving and sustaining democracy also grows. In our dependence, we have a problem: we understand how power works in the physical world, but we do not yet have a clear understanding of how power works in the digital realm.

The point of democracy is that power, when unchecked, will be abused by any person or group of people with the opportunity to wield it. More than three hundred years since the Treaty of Westphalia and over two hundred since the world's first democracy was established, we have a reasonable understanding of how nation-states wield power and how state power can be constrained by constitutions, elections, laws, and international treaties.

We lack a similar understanding of how various kinds of transnational digital platforms and the organizations that build them derive and wield their power; to whom, if anyone, they are accountable; and how their power can best be constrained in a way that does minimal harm and maximum good. Many Internet and telecommunications companies have created powerful platforms for citizens to challenge governments. Yet there is no clear model for constraining these companies' power in a way that does not also diminish their value as globally interconnected spaces.

Amid all of our excitement over new technologies, our default assumption as citizens must be that governments, powerful corporations seeking market dominance, and various other interest groups will use digital networks to obtain and maintain power whenever the opportunity presents itself. If existing institutions and mechanisms are inadequate to constrain the abuse of power across globally interconnected digital networks, political innovation must catch up to technological innovation.

Democracy was never advanced by people asking politely. Building and preserving democracy in the physical world remains a constant struggle. Conceiving and implementing a governance structure for a globally interconnected digital world that constrains the abuse of both government and corporate power and protects individual rights—based on the notion of consent of the networked—is not going to be any easier.

A first step in that direction is to acknowledge, protect, and nurture a powerful counterweight to government and corporate power: the digital commons.

CHAPTER 2

Rise of the Digital Commons

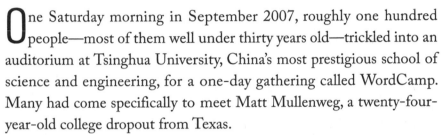

WHERE
EVERYONE
GATHER
THIS CROSS
MIND-

One Saturday morning in September 2007, roughly one hundred people—most of them well under thirty years old—trickled into an auditorium at Tsinghua University, China's most prestigious school of science and engineering, for a one-day gathering called WordCamp. Many had come specifically to meet Matt Mullenweg, a twenty-four-year-old college dropout from Texas.

In 2003 Mullenweg launched an open-source blog publishing software program called WordPress. Unlike commercial blogging software offered by companies, WordPress welcomed anybody with software programming skills to modify its software code. In that way, people could shape WordPress to fit their own specific needs, which the original creators were unable to anticipate. The free software quickly formed a global community of developers who expanded its features and functionality, so that it could be used for a range of online publishing purposes. It has been translated into a range of languages. It powers a vast range of sites from Global Voices to the Kenyan citizen media website Ushahidi, to several Arabic blogging community websites. It also powers the blogs of some of China's most cutting-edge bloggers.

ALLOWING
THE INSPIRED
PEOPLE
TO CREATE
SOMETHING.

One of the WordPress fans who clamored for a photo with Mullenweg at Beijing WordCamp was Zhou Shuguang, a twenty-eight-year-old vegetable seller turned blogger who writes under the pen name Zola. This "citizen journalist" has traveled the country writing about hot topics on the Chinese Internet. In 2006 he catapulted to fame as the "nailhouse blogger" whose on-site reports of a dramatic standoff between a Chongqing couple and local property developers helped break down a national media ban on the incident. In 2007 he was detained and escorted back to his home in Changsha after attempting to cover protests in the Northeast. In 2008 he went to the Sichuan earthquake zone to document the situation, then to Weng'an in Guizhou province, where there had been riots related to seething anger over abuse of power by local authorities and their relatives. He planned to go to Beijing for the Olympics, but police ordered him to stay home. As soon as the Olympics were over, he went to meet Mullenweg, his guru.

Zola runs his blog on WordPress; he has no other good publishing option. Most of China's 200 million bloggers use services hosted by Chinese companies. In exchange for the convenience of not having to worry about technical setup, these services control what their users can publish and are required by the government to censor heavily for politically sensitive content. Zola tried blogging on several of these platforms, but his posts kept getting taken down. Then he bought server space from a Chinese domestic web-hosting company and set up his own blog using WordPress, hoping to be free of control. But several Chinese web-hosting companies—also required to police the websites they host for politically sensitive content—shut down his site. Finally Zola had no choice but to find space on an overseas web host and to set up his WordPress blog there. Zola has little money, so he relied on free hosting from people interested in helping him. Since WordPress software is also free, Zola was able to customize it, with help from other members of the global WordPress developer community, adding special technical features that would enable his readers to connect to his site securely, allowing people to leave comments anonymously on his site.

WordPress, and the community of bloggers who use it and improve upon it, is just one small part of a much larger phenomenon known as the "sharing economy" or the "digital commons": an important reason the Internet is so revolutionary and disruptive. A robust digital commons is vital to ensure that the power of citizens on the Internet is not ultimately overcome by the power of corporations and governments.

In the early 1800s, when the United States was still a novel political experiment, Alexis de Tocqueville observed in his classic book *Democracy in America* that the key to functional democracy was a vibrant "civil society." He described the importance of active involvement by citizens in community life: people from all social strata and professions taking personal responsibility for the safety and well-being of their communities, pushing back against perceived infringements of their rights, and sharing ideas and forming associations to solve problems and improve life for everybody.

The digital commons is the virtual equivalent of Tocqueville's civil society, through which citizens can mobilize to express their interests and protect their rights. Ideally, the digital commons—comprising primarily a collection of technical standards and free and open-source software programs, plus a range of digital media creations—can exist in a positive and symbiotic relationship with both government and the private sector. But when governments or corporations abuse their power, the commons can act as a counterweight and support network through which citizens can carve out the necessary spaces to speak and organize, and thus defend their rights and interests.

In his book *The Wealth of Networks*, Harvard law professor Yochai Benkler describes how the Internet is both the product and the incubator of "commons-based information production, of individuals and loose associations producing information in nonproprietary forms." Without their efforts, neither the Internet nor the World Wide Web—with their tremendous commercial value and political power—would exist. The digital commons is a vast and growing universe of engineering inventions, software, and digital media content, created by people who have chosen to share their creations freely, because the material barriers to and costs of organizing have dropped dramatically.

THE TECHNICAL COMMONS

Millions of coders and engineers—some working for companies to make products for sale, some collaborating in communities to create open standards and free software—have collectively built a globally networked resource used by more than two billion people. The technical standards upon which the Internet and the World Wide Web depend are at the core of the digital commons: they are not copyrighted or trademarked. They are free and open to all. The Internet's inherent value and power come from the fact that it is globally interoperable and decentralized, so that everybody can add to the network and create products, services, and platforms on top of it without having to obtain permission or license, or some kind of access code, from anybody in particular. Some people choose to make businesses from their creations, others choose to share them freely, benefiting in nonmonetary ways from doing so.

The technical protocols that enable different kinds of computers and software to communicate with one another across networks are called TCP/IP (Transmission-Control Protocol/Internet Protocol). These protocols were developed by engineers working across a variety of research labs and companies. They agreed to share the protocols openly to facilitate the Internet's broader adoption and use. Imagine what would have happened if they had instead slapped a patent on their invention and tried to charge licensing fees to anybody who wanted to incorporate TCP/IP standards into their computers, software, or networking devices. The Internet would not be what it is today if those engineers had been unable or unwilling to share TCP/IP freely with the world as part of the digital commons.

The World Wide Web, invented two decades after the Internet, similarly owes its existence to the digital commons. What we call "the web" is an interlinked universe of websites accessed through web browsers such as Internet Explorer, Safari, and Firefox. All websites, no matter where they are created or what kind of computer system is used to host them, are readable from anywhere thanks to a common computer language called the hypertext mark-up language (HTML). The web is how most people on the planet today use the Internet. We can thank the Englishman Sir Tim Berners-Lee (eventually knighted for his invention), who in 1990 while at the particle physics lab at the European Organization for Nuclear Research in Switzerland wrote a simple computer program called the WorldWideWeb to make it easier for researchers in his lab to locate and share each other's data. Sir Berners-Lee did not try to patent or charge for the use of his HTML language and the web-addressing system he created; instead he released them into the public domain. The Web would not be truly worldwide had he sought to license the use of the language that enables people to create websites.

Even though the digital commons is based on protocols and technologies that were developed noncommercially, Internet companies like Google have had a symbiotic and overlapping relationship with the digital commons from the beginning. None of these businesses would be what they are today if it hadn't been for the work that network engineers and computer scientists contributed to the commons. For this reason, most large Internet companies—from Cisco, AT&T, and Intel to Microsoft, Google, and Facebook—support and participate in the work of open standards organizations (such as the Internet Engineering Task Force, or IETF, and the World Wide Web Consortium, or W3C) and the coordinating bodies that maintain the global addressing and domain name system (like the Internet Corporation for Assigned Names and Numbers, or ICANN, and regional Internet registries). These organizations maintain, update, and add to the technical standards necessary to keep the Internet running on increasingly sophisticated devices used by almost two billion people around the planet. Governments do not run any of these organizations.

Some countries—China, Iran, and Brazil, to name a few—have lobbied for the United Nations to take over these organizations and coordinating bodies. Other countries, including the United States, are opposed. So are human rights groups and most major Western Internet companies; Google has taken a particularly public stand on this issue. Most Internet users around the world are unaware of this increasingly high-stakes fight over the future of Internet governance, a geopolitical battle with implications for the future political freedoms of all citizens.

The free-software and open-source software communities are also key to the expansion and growth of the digital commons, upon which a great deal of both commercial and noncommercial activity now depends. In 1989, the computer scientist Richard Stallman got the ball rolling when he created the General Public License (GPL), which authorizes anybody to use a GPL-licensed software program as long as any copies or derivatives are also made available on the same terms. This license enabled software programmers to contribute computer code with the express purpose of sharing it with others, who can in turn build and improve on it, on the condition that the modifications and improvements remain part of the commons.

GPL software licensing was brought into the global prime time by Finnish computer scientist Linus Torvalds, who created the "kernel" of the Linux computer operating system, which programmers all over the world have worked to build upon and improve. In 2000, IBM adopted Linux as the centerpiece of its service and consulting business. By 2008, 60 percent of the world's computer servers ran on Linux, with a minority 40 percent using Microsoft's proprietary Windows platform. Open-source web server software called Apache runs more web servers in the world than any other, beating out all commercial alternatives. Open-source website design and management software like Drupal and WordPress—developed by a global community of programmers— run the websites of millions of small businesses as well as millions more blogs and nonprofit websites. Human rights groups, small news organizations, community media groups, and activists around the world are heavy users of open-source software. Many of the digital activists who helped bring down the governments of Egypt and Tunisia were heavy users of—and in some cases contributors to—open-source software.

Open-source software is not inherently antibusiness. Some open-source projects and communities work separately or even in direct competition with companies, but many work in a symbiotic relationship with for-profit businesses. Google heavily uses and contributes to the digital commons, and so does Facebook. The Android mobile operating system developed by Google runs on Linux, and its openness is one of the key reasons Android has quickly become a key driver of the global smart-phone market. According to the Linux Foundation's 2010 annual report, since 2005 around 6,100 individual developers and six hundred companies have contributed to the Linux kernel.

In December 2010, Reuters News Agency reported that more than 70 percent of contributions made to the Linux code base "are from developers who are getting paid for their Linux development from corporations who hope to benefit from better software in their core business." In a 2009 speech, Jim Zemlin, executive director of the Linux Foundation, pointed out that every Internet user on the planet is a Linux user in some way or another, because there is almost no digital product or service today that does not rely upon or in some way benefit from the open-source operating system.

ACTIVISM

The digital activists who played a key role in bringing down the governments of Tunisia and Egypt did not spring immaculately from Facebook and Twitter. They started building interconnected online and offline movements well before either service was conceived, using a variety of platforms, services, and software—some built by companies, some built by themselves or created noncommercially by others. They were also major contributors to—and beneficiaries of—the digital commons.

Riadh Guerfali and Slim Amamou, plus a number of others including Sami Ben Gharbia, a Tunisian dissident exiled in the Netherlands, started experimenting with online activism back in 2000. Their first project was an online magazine on which they posted articles e-mailed to them by less tech-savvy Tunisian dissidents and activists who were banned from publishing in Tunisia and had no other way to get their information disseminated. In 2004 they used WordPress to create Nawaat.org, a website that later won awards for its role as an information hub for activism during the "jasmine revolution" of late 2010 and 2011.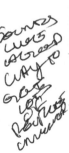

For the next five years, Nawaat's importance grew as a hub for videos, photos, and articles about the human rights situation in Tunisia, the latest outrages of Ben Ali's regime, information and updates on what was being censored and how, plus tutorials about how to get around censorship and evade surveillance. Nawaat's material was published under a Creative Commons (CC) license, a more open form of copyright licensing—inspired by the GPL open-source software licensing system—which allows content to be shared and copied freely as long as credit is given to the original source. This type of licensing is particularly useful for activists and civic organizations trying to reach wide audiences and whose primary purpose is not financial profit. CC licenses encourage people to copy and republish activist content without worrying about being accused of stealing or pirating content that was, after all, created with the express purpose of being shared and viewed as widely as possible.

As new start-up companies in Silicon Valley and elsewhere spawned new social media tools, the Nawaat team quickly adapted them for activism. Not long after Google launched Google Maps, for example, Sami Ben Gharbia used it to create the Tunisian Prison Map, calling attention in a dramatic and visual way to just how many political prisoners were being held around the country. Though much of Nawaat's content is republished on YouTube, Flickr, and Facebook and is promoted heavily on Twitter, the activists learned early how important it would be to keep all of their content archived on their own platform, over which they have complete control. At one point Nawaat's YouTube channel was frozen by company administrators because somebody—nobody knows who—filed a complaint through YouTube's internal abuse-reporting system that some of the channel's content violated the company's terms of service. Similar complaints were frequently filed against the accounts of Nawaat team members on Facebook, because its terms of service require that people use their real names on their account—something that was too risky for most Tunisian dissidents, who preferred to use pseudonyms in hopes of not becoming another data point on the Tunisian Prison Map.

Over the years, Tunisian activists developed close ties with bloggers and activists around the Arabic-speaking world, and particularly bloggers from Egypt, which happened to be home to a highly educated, early-adopting techie community. By 2004, Egyptian bloggers were collecting and disseminating information about arrested activists, as well as shocking acts of torture by the Mubarak regime's police forces. A number of these bloggers were arrested in connection with offline street protests as well as online writings. When arrests took place, other bloggers used all digital platforms available to raise public awareness about what had happened to them and to circulate information about their arrests and activities throughout the Arabic-language blogosphere.

One of many activists who spent time in jail during those early days of online activism was Alaa Abd El-Fattah, who in 2004 with his wife, Manal, also a blogger, used the open-source publishing software Drupal to launch a website aggregating the posts of Egypt's political bloggers. The Egyptian Blogs Aggregator helped to build a community of people who then—like the Tunisians—adopted whatever social networking and web tools best served their purposes. By 2006 Facebook had gained a critical mass among young Egyptian Internet users. A group of young activists found it to be particularly useful in organizing street protests, as well as to call attention to the arrests of their friends. The rest is history. In 2010 after the death of a young man named Khaled Said at the hands of Egyptian police, Google executive Wael Ghonim, using a false identity, created the Facebook page "We are Khaled Said," which became a hub for organizing antitorture protests in 2010 and organized the first protest in Tahrir Square on January 25, 2011, after which events snowballed.

Before the Arab Spring of 2011, during which protest movements organized in part with the help of the Internet successfully toppled the leaders of Tunisia and Egypt and inspired revolts around the region, Tunisian and Egyptian bloggers played leading roles in organizing two pan-Arab blogger conferences. The Arab Techies Collective, formed in 2008, brought together a more technical group of software developers and web designers from around the region, who collaborated on open-source software development projects. They worked together to modify, improve, and customize open-source software that they found particularly useful for Arabic-language activism, translating software interfaces as well as manuals into Arabic. The Arab techies also developed ties with developer communities in the United States, Europe, and Asia. These online and offline ties were put to good use as tensions began to rise in Tunisia and then in Egypt. As websites around the region were blocked and hacked by governments seeking to thwart their movement, activists already knew whom to contact in a global community of developers who work on open-source anticensorship and anonymization software. An international community of "hacktivists" quickly reached out to help.

Given their preexisting ties with activist hackers around the world, it was not surprising that the Tunisian activists running Nawaat were able to get ahold of WikiLeaks' trove of US diplomatic cables pertaining to the Ben Ali regime. The revelations posted on the website they created, Tunileaks, confirmed to the Tunisian public that US diplomats viewed the Ben Ali regime as corrupt and increasingly dysfunctional, despite not saying so publicly due to Ben Ali's perceived helpfulness in the war on terror. The information published by Nawaat and Tunileaks is believed by diplomats, human rights groups, and journalists who have since gone back and analyzed the Tunisian revolution to have contributed in no small part to Tunisians' rejection of the Ben Ali regime. This explanation rings true given the timing of the information's release, as street protests mounted in the wake of the self-immolation of a desperate vegetable seller in the town of Sidi Bouzid.

The emergence of the Arab digital commons and its natural relationship with the broader global commons is a textbook example of what Clay Shirky, a technology and social theorist at New York University, has described in two influential books about how people use and contribute to the commons. The Internet, he argues, makes it easier than ever for the civic-minded to rally around a cause and to push for change. Wikipedia, the all-volunteer, peer-produced online encyclopedia, was perhaps one of the earliest and most dramatic examples of this phenomenon. Blogging, tweeting, podcasting, and taking pictures and videos and uploading them to Flickr and YouTube can all be done at near-zero financial cost.

Such "social production" is not exclusive to any side of the political spectrum or to a particular set of values. Viral homemade videos on YouTube helped mobilize young people to volunteer on the campaign trail and vote for Barack Obama in 2008. Similar tools and activities were key to the Tea Party–led Republican takeover of the US House of Representatives in 2010. WikiLeaks, the whistle-blowing website most famous for its controversial release of leaked US diplomatic cables, is also part of the digital commons.

BALANCE OF POWER

Though the digital commons played a vital part in facilitating the Arab Spring, it is less clear how it will contribute to Arab democracy. What *is* clear, however, is that software code and technical infrastructure have an important role in mediating the relationships between citizens of these countries and their new governments—alongside laws, constitutions, institutions, and political processes.

As Harvard law professor Lawrence Lessig explained more than a decade ago in his seminal book, *Code and Other Laws of Cyberspace*, software code and technical standards are for all practical purposes a new form of law, because just like laws, they shape what people can and cannot do. The implications are massive. In the pre-Internet era, government—which in democracies at least is expected to reflect "consent of the governed" and to be held accountable to the public interest—had the primary responsibility for developing legal codes governing what people did in the physical world, backed up by the authority and force necessary to enforce meaningful levels of compliance. In the Internet age, a whole new sphere of de facto lawmaking has emerged in the guise of software code and technical standards that channel and constrain what people do with their technology. In this way, the power to shape how citizens can organize, access and disseminate information, and express their own ideas and opinions has expanded dramatically to anybody who knows how to write software code, create Internet and telecommunications hardware, and build interoperable networks. Right now this power is wielded primarily by a combination of people who are either working in service of companies or contributing to the commons in some fashion.

[handwritten margin note: Citizens of Networks?]

People who contribute to the digital commons and who act as its stewards, architects, and defenders are engaging in a form of citizenship for the Internet. Some now call this "netizenship." David Bollier, whose book *Viral Spiral* documents the rise of the digital commons in the United States, argues that the people who build and contribute to it—the "commoners"—have "invented a new sort of democratic polity." Many American and European technologists assume in their writings that the emergent digital commons is inherently more democratic than any existing parliamentary democracy, because governments and laws tend to lag hopelessly behind technological change. But is it logical to assume that a new digital civil society, whose vanguard is led by computer programmers, tech-savvy early adopters belonging to a range of groups, hackers, and bloggers with a range of political and economic agendas, will necessarily serve the interests and respect the rights of all people who use the Internet across the globe? Who has the right to impose their ideas of right and wrong in shaping the network's future?

Despite the optimism of Bollier and many Western technologists, unless human nature undergoes some sort of fundamental transformation, it seems inconceivable that the world's "netizens" will naturally act, in aggregate, in a way that serves the common good and respects the rights of all vulnerable minorities and people with peaceful but unpopular views. A system is needed to reward good behavior and punish bad behavior as well as general consensus around what constitutes "good" and "bad." Plenty of crime and acts of cyber-warfare are committed through the Internet and with the help of the commons.

Different "netizens" have different political loyalties: there are pro-government bloggers and hackers in China, Iran, Russia and many other countries, proudly ready to attack their government's perceived and imagined enemies. There are jihadist programmers. There are vigilante hacker groups like Anonymous, who launched "denial of service" attacks against businesses that cut their ties with WikiLeaks, and who have been known to attack the websites of other entities whose values its members disapprove of. Though most members of Anonymous view themselves as freedom fighters protesting the unjust power of governments and corporations over people's digital lives, one can make a strong case that their chosen form of protest harms innocent people (i.e., customers and users of the services they are attacking) and is unconstructive when it comes to building a more rational and equitable set of power structures over the long term. They are more like Robin Hood, deliberately committing crimes in response to King John's governance failures, than the Levellers of the English civil war, who fought to change an elitist political structure, or the American revolutionaries who sought to build an alternative system of governance to replace an unjust one.

The need to protect the rights of unpopular minorities, the less tech-savvy, and other innocent people from criminals is one of many good reasons that the need for democratic government—governance based on consent of the governed—will not go away. Doing away with geographically based governance is not a good idea either: citizens remain at core flesh-and-blood people with physical and often very local needs. At the same time, the power of multinational corporations continues to grow, empowering and employing tens of millions while also displacing and exploiting others. It is clear to anybody who pays attention to the news that these two pillars of global society, governments and companies, cannot be counted upon alone—and even less so together—to protect the rights or serve the interests of all the world's people, in either physical space or digital space. The citizen commons—parts of which are very local and regional, and parts of which are very global—is thus a vital counterweight to government and corporate power in cyberspace. The commons provides the space, tools, and community that allow people to build noncommercial, secure, and private spaces that enable dissent, whistle-blowing, and nonmainstream conversations.

Today the existence and importance of the commons is not well recognized, valued, or understood even by many of its contributors and participants. Even in the democracies of North America and Western Europe, creators and defenders of the commons are engaged in pitched battles in legislatures and courts against powerful companies, government agencies, and interest groups who find many aspects of the Internet threatening to their businesses, security protocols, and accustomed ways of life. The weaker the democracy and the weaker its constitutional and legal protections of citizen rights, the more beleaguered the commons becomes, at citizens' expense.

At the extreme end of this spectrum—that is, the spectrum of governments that have learned to actively manipulate technology rather than simply restrict citizens' access to technology, as old-fashioned authoritarian regimes like Cuba and North Korea currently do—is China. The robust, lively, and rapidly growing Chinese Internet is the product of a strong government plus a robust private sector, combined with the absence of democracy, accountable governance, due process, and meaningful protection of citizen rights. China represents a new, upgraded form of the authoritarian state: it embraces Internet connectivity not merely as essential for a world-class economic and financial power, but also as necessary for modern government. At the same time, China's digital commons is under attack and subject to cynical manipulation by the state, with the direct assistance and collusion of the private sector.

PART TWO

CONTROL 2.0

We will never have a real civil society, a democratic society, unless people take responsibility.

—AI WEIWEI, JANUARY 2009

CHAPTER 3

Networked Authoritarianism

In 2009 in Hubei province, a twenty-one-year-old waitress named Deng Yujiao was working as a manicurist for a business that combined the services of restaurant, spa, and salon. Patrons could wine and dine, then relax with a haircut, manicure, and massage without leaving the building; sexual services were also known to be available. A local Communist Party official, entertaining friends at this multiservice establishment, took a shine to Deng. He demanded sex; she refused. He tried to force her. She fought back with a pedicure knife and somehow ended up killing him.

News of the incident quickly spread online. Deng was hailed as a heroine. A blogger tracked down the local mental hospital where she was being held and posted pictures online. Women wrote passionate postings titled "We are all Deng Yujiao." Despite government efforts to control the news, people were simply too angry—about abuse of power by petty local officials, as well as about the economic circumstances that compel young women to make a living in such sleazy establishments. Realizing that a conviction could spark riots, the authorities eventually dropped the murder charges against her. She was convicted on a lesser charge instead and set free.

In the pre-Internet age, such a person would have disappeared into the prison system or into a mental health ward, unbeknownst to anybody other than a few close friends and relations in Hubei. The Internet enables ordinary Chinese people to speak truth to power and pursue

[handwritten margin note: EVERY-THING WOULD BE POSTED ONLINE —]

justice in unprecedented ways. At the same time, Chinese Internet users have a manipulated and distorted view of their own country as well as of the broader world. *THEY ARE NOT ALLOWED TO USE THE INTERNET THATS WHY.*

I had a telling encounter on a recent visit to Peking University. Once a hotbed of political activism during the 1989 student protests, these days the university keeps students under considerably tighter control. I met a doe-eyed undergraduate dressed in stylish jeans and a fuzzy sweater who wanted to interview me for her student magazine. (I will call her "Mary" to spare her trouble and possible embarrassment.) She asked about my experiences covering China as a CNN correspondent and how the Chinese government handled the Western media. I told her my colleagues and I frequently had been detained by police in the 1990s.

HAVE U WOTEN THE BOOK MAN

"Really? I didn't know that happened to foreign journalists," she said.

"It still does," I replied. "You know of course that Chinese journalists sometimes go to jail if they push too far with their investigative reporting."

Silence. Mary's large black eyes, framed by tasteful amounts of eyeliner and mascara, widened and began to glisten as she stared at me in shock. "I had no idea," she said softly.

In all modern societies, there are heated disputes over how media and information networks are regulated and managed, and the extent to which systemic biases favor people with power and money over those without power or money. But in China there is no transparency or public accountability when it comes to how information networks are shaped, operated, regulated, and policed. This total opacity, plus government co-optation of the private sector in carrying out political censorship and surveillance, are the key components of what I call China's networked authoritarianism. The key to the system's success is that the regime does not try to control everybody all the time; its controls on political information are nonetheless effective *enough* that most Chinese—including educated elites—are unaware, or have a distorted view, of many issues and events in their own country, let alone in the rest of the world.

Good ADVICE for CHINA AND ANYONE PLACE ALIVE.

In 2005 a PBS documentary crew visited university campuses around Beijing and showed students the iconic 1989 photograph of a man standing in front of a tank. Amazingly (at least to Westerners), most didn't recognize the image at all. Chinese university students' blinkered knowledge and understanding of their own society, past and present, is one of the regime's most deliberate and profound achievements.

The 1989 Tiananmen Square protests were sparked by the death of a reformist leader, Hu Yaobang, but the fire was fueled by the historic Beijing visit of Soviet leader Mikhail Gorbachev soon thereafter. Students hoped that Chinese leaders would follow his policy of glasnost, the Russian word for "openness," which became the catchphrase for a loosening of controls on the Russian press and discussion of political reform. After the bloody June 4 crackdown, Deng Xiaoping quashed all hopes that China would follow Gorbachev's lead. Instead Deng focused aggressively and exclusively on the Chinese version of perestroika, or "restructuring": accelerating the economic reforms that have made China the world's second-largest economy today and a rising global power. Deng was betting that most Chinese people would agree that the trade-off between economic prosperity and political liberty was worthwhile—or at least enough of China's elites would believe so and maintain their allegiance to the Communist Party as the guarantor of prosperity and stability.

Deng's efforts have nevertheless given way to a limited sort of cyberglasnost: people use the Internet to address injustice at the local and personal level, expanding the public discourse and enabling people to speak out on a greater range of issues, while at the same time no meaningful changes have been made to the political system. People who cross the government's political red lines find themselves with no better legal protection than they had twenty years ago. These people include 2010 Nobel Peace Prize winner Liu Xiaobo, who was arrested in late 2008 and sentenced in 2009 to ten years in jail for state subversion for writing the prodemocracy treatise Charter 08. In early 2011 the artist and political activist Ai Weiwei—internationally famous for his unorthodox art installations as well as his outspoken views on corruption and political reform—was detained and investigated for tax fraud, although it was clear that the real reason for his detention was that he repeatedly linked cases of local corruption and malfeasance with the need for broad democratic reforms.

According to the Dui Hua Foundation, a China-focused human rights research and advocacy organization, in 2008 (the year of the Beijing Olympics) the number of people sentenced for "endangering state security"—the most common charge used in cases of political, religious, or ethnic dissent—more than doubled for the second time in three years. Since then, political arrests have remained substantially higher than in the 1990s, when few Chinese people had Internet access. Thanks to censorship, the average person rarely encounters information about these types of arrests.

A networked authoritarian regime benefits from the lively online discussion of many social issues and even policy problems. The government follows online chatter, which alerts officials to potential unrest, better enabling authorities to address issues and problems before they get out of control. Sometimes people are successful in using the Internet to call attention to social injustice, as in Deng Yujiao's case. Citizens even manage to use cyber-vigilantism—popularly known as the "human flesh search engine"—to bring about the resignation of corrupt officials. Sometimes laws or regulations are even changed as the result of concerned citizens' online campaigning. As a result, China's 500 million Internet users feel more free and are less fearful of their government than in the past. The communications revolution has transformed China in many ways, but the Communist Party has thus far succeeded—against all odds and expectations—in controlling it to remain in power.

HOW CHINA'S CENSORSHIP WORKS

In fall 2009, I sat in a large auditorium festooned with red banners and watched as Robin Li, CEO of Baidu, China's dominant search engine, paraded onstage with executives from nineteen other companies to receive the China Internet Self-Discipline Award. Officials from the quasi-governmental Internet Society of China praised them for fostering "harmonious and healthy Internet development." In the Chinese regulatory context, "healthy" is a euphemism for "porn-free" and "crime-free." "Harmonious" implies prevention of activity that would provoke social or political disharmony.

China's censorship system is complex and multilayered. The outer layer is generally known as the "great firewall" of China, through which hundreds of thousands of websites are blocked from view on the Chinese Internet. What this system means in practice is that when one goes online from an ordinary commercial Internet connection inside China and tries to visit a website such as http://hrw.org, the website belonging to Human Rights Watch, the web browser shows an error message saying, "This page cannot be found." This blocking is easily accomplished because the global Internet connects to the Chinese Internet through only eight "gateways," which are easily "filtered." At each gateway, as well as among all the different Internet service providers within China, Internet routers—the devices that move the data back and forth between different computer networks—are all configured to block long lists of website addresses and politically sensitive keywords.

These blocks can be circumvented by people who know how to use anticensorship software tools. It is impossible to conduct accurate usage surveys, but it is believed likely that hundreds of thousands of Chinese Internet users deploy these tools to access Twitter and Facebook every day. Yet researchers estimate that out of China's 500 million Internet users, only about 1 percent or so (a number somewhere in the single-digit millions—still a large number of people but not enough percentage-wise to shape majority public opinion) use these tools to get around censorship, either because most do not know how or because they lack sufficient interest in—or awareness of—what exists on the other side of the "great firewall."

Fortunately for the government, there are plenty of social networking platforms and other delightfully entertaining and useful services on the Chinese Internet to keep people occupied, without much need to access sites and services based overseas—assuming they have no interest in politics, religion, or human rights issues. Baidu, the homegrown search engine, enables people to locate all the content on the Chinese-language Internet that their government permits. The social networking platforms RenRen and Kaixinwang substitute for Facebook. People can blog on platforms run by Chinese companies like Sohu and Sina, which also runs a wildly popular Twitter-like microblogging service, Weibo. QQ, run by the company Tencent, offers instant messaging, gaming, and all kinds of interactive services that work seamlessly across both PCs and mobile phones. These companies have all benefited from substantial Silicon Valley investment over the past decade, and many are listed on US stock exchanges or others outside of China. Thanks to the many Americans who find China's rapidly growing Internet market to be an irresistible investment opportunity, these companies are well funded to provide highly entertaining and useful—albeit censored and heavily monitored—content and services.

These domestic companies are the stewards and handmaidens, the tools and enforcers, of China's inner layer of Internet censorship. Why simply block content when you can delete it from the Internet for good? Why hire government employees to carry out censorship and surveillance when companies can be compelled to do it? The government requires companies operating inside China to use a combination of computer algorithms as well as human editors to identify objectionable material and remove it from the Internet completely. Companies that fail to obey government orders face different grades of punishment: from warnings or stiff fines to temporary shutdowns or revocation of the company's business license. Many thousands of Chinese websites and dozens of companies have been shuttered because they failed to control their content adequately.

This requirement of corporate self-censorship applies to all Chinese websites, from small online communities to the largest commercial sites, like Baidu. It also applies to all foreign Internet companies with operations inside China—including Google.cn before Google decided to pull out. Google's experience with Chinese censorship helps illustrate how these different layers work. Before it entered China in 2006, Google operated outside the "great firewall," which means that it was subject to blocking by the Chinese network. For example, if one were on the Internet in China and typed the Chinese characters for something politically uncontroversial, say, "automobile," into the search box on Google.com, everything would work fine. But if you tried to search for anything politically sensitive—such as the Chinese phrase for "Tiananmen Square massacre" or something related to politically sensitive breaking news, like the name of a city where a riot had just occurred—the page would be blocked. The page existed on a server overseas, but it could not be viewed in China. In other words, the search was blocked not by Google but by Chinese network engineers.

Then in 2006, Google decided that subjecting users to the inconvenience and frustration of such increasingly frequent blockages was not the best way to attract Chinese Internet users to its search engine. So they launched Google.cn inside China, agreeing to abide by the Chinese government's censorship requirements. To gain permission to operate from within the firewall, Google had to agree to adjust its search algorithms so that results on Google.cn would not include websites blacklisted by the Chinese government. Rather than get a blank page when searching for the Chinese name of a city where a riot had recently been put down by police, users of Google.cn would get a sanitized set of search results about that city, minus web pages containing reports from human rights and dissident websites.

After Google announced in January 2010 it was reconsidering its business in China, then pulled its search engine out of China in March, the government imposed strict media controls on the story. As a first line of defense, the "great firewall" blocked overseas Chinese-language news reports about Google's decision to remove its search engine. The government also deployed a range of offensive tactics: all blog-hosting services, microblog platforms, and social networking services operating inside China were required to censor what Chinese Internet users said about Google. Authorities issued specific instructions to spin and manipulate the domestic media, in an aggressive effort to shape public opinion about what had happened. Not that people could not say anything: they were free to show Google in a negative light, and there were plenty of Chinese Internet users happy to trash Google, as there are all over the world.

But Chinese bloggers and social network users who expressed sympathy for Google's situation quickly found their postings deleted and blocked by all Internet companies—domestic and foreign—operating inside China. Writings by liberal-leaning people who argued that the free flow of information would be better for China's economy and that censorship only makes it harder for the Chinese government and people to resolve problems were also deleted. The government's State Council Information Office issued a direct and detailed order on the subject to all websites and news organizations. A Chinese blogger obtained the full text and posted it online. Here is a portion of that text, translated by the California-based website *China Digital Times*, run by the exiled activist Xiao Qiang:

1. It is not permitted to hold discussions or investigations on the Google topic.
2. Interactive sections do not recommend this topic, do not place this topic and related comments at the top.
3. All websites please clean up text, images and sound and videos which attack the Party, State, government agencies, Internet policies with the excuse of this event.
4. All websites please clean up text, images and sound and videos which support Google, dedicate flowers to Google, ask Google to stay, cheer for Google and others have a different tune from government policy.
5. On topics related to Google, carefully manage the information in exchanges, comments and other interactive sessions.
6. Chief managers in different regions please assign specific manpower to monitor Google-related information; if there is information about mass incidents [the Chinese euphemism for "protests"], please report it in a timely manner.

Such directives are common, forcing Internet companies to maintain entire departments full of people whose job it is to respond to them. In late December 2010, Wang Chen, deputy head of the Communist Party's propaganda department and chief of the State Council Information Office—two of several party and government bodies in charge of Internet censorship policies—boasted in a speech that 350 million pieces of "harmful content" had been deleted from the Chinese Internet over the course of one year. Earlier that year, in a presentation to top government leaders, Wang gave a detailed description of an Internet "management system that integrates legal regulation, administrative supervision, industry self-regulation, and technological safeguards." Some sections of his speech (the full text of which was leaked to the New York–based group Human Rights in China) were deleted from the publicly released version. One of these deleted sections, which the government did not intend to share with the public, said:

We are following the overall thinking of combining Internet content management with industry management and security supervision; combining prior review and approval with supervision afterwards; combining technological blocking with public opinion guidance; combining hierarchical management with local management; combining government management with industry self-regulation; and combining online monitoring with offline management.

In such an environment where search engines and social networking services are so heavily censored, most people are not even aware of the existence of many facts, incidents, or ideas unless somebody they know who is technically savvy enough to access uncensored online spaces happens to e-mail a link to them. People who use domestic e-mail services and social networking platforms to disseminate such information, of course, are subject to monitoring and potential arrest. Data-mining software and "deep packet inspection" technologies make it easy to automate surveillance through the Internet service providers and mobile carriers of all unencrypted Internet traffic no matter what service is being used or where it is based.

In 2011, the government moved to extend these censorship and surveillance mechanisms, as well as to improve their coordination. In March 2011, spooked by the Arab Spring, the central government established a new overarching government agency responsible for controlling all Internet platforms and services. The number of censored foreign websites, social networking platforms, and even data-hosting and "cloud computing" services expanded dramatically. Surveillance systems were upgraded to more aggressively track and identify Chinese citizens who managed to circumvent the blockages to use tools like Twitter. It became commonplace for Twitter users to be questioned about their postings, and at least one person was arrested for no other reason than a tweet she had sent out.

AUTHORITARIAN DELIBERATION

At 5 a.m. July 16, 2009, Guo Baofeng, a young entrepreneur who lives in the city of Mawei in the southern coastal province of Fujian, posted two short messages in English onto Twitter from his Blackberry:

Pls help me, I grasp the phone during police sleep
I have been arrested by Mawei police, SOS

Guo's tweets, sent furtively before police confiscated his phone, were quickly retweeted by hundreds of people in China and around the world. Nothing more was heard from him, but those short tweets were enough. People in his Twitter network who knew his real-life identity followed up with his family and friends in Mawei. News quickly spread around Twitter that the police had taken Guo from his office the previous afternoon.

Along with several other bloggers, Guo was arrested on defamation charges, after posting information online about the alleged gang rape and murder by local officials of a young woman named Yan Xiaoling. News of Yan's case had spread virally in chat rooms and blogs after the young woman's parents insisted she was violently raped, contradicting police claims that she died from a hemorrhage caused by an ectopic pregnancy. Authorities scrambled to stop the further spread of allegations online. The Communist Party's propaganda department instructed news outlets and social media websites to delete all discussion of Yan's case, but some people like Guo continued passing around the news on foreign social media platforms like Twitter to reach those few hundred thousand Chinese users who know how to circumvent the firewall.

Guo's Twitter followers immediately began to rally for his release. One well-known Guangzhou-based blogger named Wen Yunchao called for a postcard-writing campaign to shame authorities into releasing him. Hundreds of people mailed postcards to the Fuzhou detention center where Guo was being held, with the simple message: "Guo Baofeng, your mother wants you home for dinner." Others organized a fund-raising drive to pay for his defense. After sixteen days in detention, Guo and two other bloggers who had been arrested around the same time were released. "I used Twitter to save myself," he wrote on his blog. The massive online reaction, he believes, helped to free him.

The people who rallied to free Guo did not spring immaculately from cyberspace—or from Twitter. Many of them had gotten to know one another both online and offline over the course of several years through a loose community of liberal-thinking, internationalist-minded bloggers, web developers, and entrepreneurs. For five years beginning in 2005 they even held annual conferences. They were, in essence, the reformist, prodemocracy branch of China's digital commons. In November 2009, the Sichuan-based writer Ran Yunfei stood in front of several dozen liberal Chinese bloggers at the fifth annual Chinese Blogger Conference, held in the small town of Lianzhou, Guangdong province, and explained how vital it is that people continue speaking their minds online. "As we use the Internet every day, it changes us. It has made me more tolerant and taught me to play by a set of rules. . . . As we train ourselves we are also training the government. . . . Hopefully one day they will understand that they don't need to be afraid of us, that we can all legally and rationally coexist."

The Chinese Communist Party acted aggressively in 2010 and 2011 to smash all dreams of any such coexistence. When the same group of bloggers tried to organize another gathering in November 2010, police threatened organizers, and what would have been the sixth annual Chinese Blogger Conference was canceled. In early 2011, after the Tunisian and Egyptian governments fell to popular uprisings that came to be known as the "jasmine revolution," calls for "jasmine" protests in China began to circulate in the Chinese twittersphere. Ran Yunfei was arrested and charged with subversion, although he appeared to have done nothing more than to retweet some controversial messages about events in Tunisia and Egypt, with comparisons to China. Other members of the liberal Chinese blogger community—many of whom had also signed Liu Xiaobo's Charter 08 two years previously—were detained for days or weeks. Many more were questioned by the police and warned that they were being closely watched. Transcripts of people's posts on Twitter and on other social media platforms featured in interrogation sessions. China's liberal reformist commons was quashed. Thanks to censorship, most educated Chinese people never knew this community existed.

Herein lies the paradox of the Chinese Internet: public debate and even some forms of activism are expanding on it, while at the same time, state controls and manipulation tactics have managed to prevent democracy movements from gaining meaningful traction. Min Jiang, a scholar of the Chinese Internet at the University of North Carolina at Charlotte, calls this sort of lively but limited public discourse "authoritarian deliberation." Her argument centers around the idea—often overlooked by Western punditry—that it is possible for a large amount of public debate, and even some kinds of policy activism, to exist within the confines of an authoritarian state. A major difference between democratic and authoritarian deliberation is that in the authoritarian context, the state sets limits to political speech. The state conducts surveillance against which citizens have no legal recourse. The right to free expression is not guaranteed and cannot be defended in court because the judicial system is not independent. But thanks to the Internet, people have vastly more room to voice their concerns and criticize public authorities than ever before—even in China.

Chinese government officials and the state-run media actually talk about Chinese-style "Internet democracy" as an alternative to "Western-style" multiparty democracy. China's parliament now has an "e-parliament" website, on which it invites members of the public to make policy suggestions. Outsiders who do not read Chinese or follow China's Internet scene closely would be very surprised to learn about some of the suggestions people are allowed to post. People post detailed suggestions on a range of subjects, such as reducing local corruption, improving the environment, and proposing financial reforms. In 2009, the first year the e-parliament was launched, somebody posted a long treatise about why China's one-child population-control policy should be scrapped. In the comments thread following the posting, a long argument ensued among people from across the country, some supporting the policy's abolition, others defending it as a key to China's economic success, and still others accepting some kind of population-control policy but proposing that the current one be made more flexible.

As it happens, for the past several years the central government has been modifying China's one-child policy as the population ages and the birth rate drops, making such conversations useful in gauging potential public reactions to policy changes. At the same time, attempts to post proposals for multiparty electoral reform on the e-parliament website will be censored. But the government nonetheless points to such websites to demonstrate that it is listening to public opinion and taking it seriously. They even quote President Hu Jintao, who gave his blessing to the whole enterprise back in 2008 when he said to the *People's Daily*, "The web is an important channel for us to understand the concerns of the public and assemble the wisdom of the public."

Since President Hu made those remarks and conducted his first public web chat, Chinese leaders at all levels have been holding regular online chats with the public. Officials at all levels of government—even police chiefs in some cities—have set up public accounts on Weibo, the

Chinese version of Twitter, to demonstrate an interest in public opinion as well as to help the public better understand their priorities. Of course, the various online platforms through which the public is invited to converse with officials are policed and censored. Postings that might contain serious critiques of the whole system, or that call for political reform, are eliminated.

Thus the Chinese government can maintain an authoritarian state, while at the same time the Internet enables a high level of lively and even contentious online debate and deliberation, within certain limits. After a July 2011 high-speed-rail crash in the city of Wenzhou that killed at least forty people and injured nearly two hundred, Chinese Internet users were able to use Weibo to tweet a stream of on-the-scene reports as well as critical invective about corruption and malfeasance that contributed to the crash, without the Communist Party's power being fundamentally threatened. After an outburst of online rage in the immediate aftermath of the crash, the government moved to muzzle all professional media coverage and also tightened the screws on Internet companies and social networking companies to track and censor critical postings, especially any that linked the crash to broader national governance failures or attempted to organize offline responses. As a result, the spontaneous outpouring of anger—which was too big and sudden to be contained—that followed the crash was prevented from morphing into offline protests or any broader movement for political change.

Meanwhile, as the government has moved aggressively to quash the communities in the Chinese digital commons harboring people who are sympathetic to calls for political reform, government-friendly communities are allowed to thrive. Some are the product of direct manipulation but others are largely spontaneous and voluntary. To discredit liberal viewpoints in Chinese cyberspace, the government deploys other means to sway public opinion by—to use an American political and commercial marketing term—"Astroturfing" the Chinese Internet with progovernment commentary. Communist Party youth league volunteers and paid freelance recruits known derisively as the "fifty-cent party" (because they are paid roughly fifty Chinese cents per posting) lurk on blogs and chat rooms, posting patriotic and progovernment comments and reporting on excessive political deviance. But beyond the Astroturfers, there are also many young patriotic Chinese who create online communities on their own initiative. One example is April Media, an online community of nationalist bloggers run by a young man named Rao Jin, a recent graduate of the prestigious Tsinghua University in Beijing, dedicated to exposing the Western media's "anti-China bias."

Rao launched the site, originally titled Anti-CNN, in March 2008 in the wake of ethnic unrest and the subsequent Chinese government crackdown in Tibet. Western media coverage of the news included highly critical commentary of the Chinese government, including a comment by CNN commentator Jack Cafferty calling them "goons and thugs." (Cafferty and CNN later apologized.) Editors of major Western TV broadcasters also made serious errors in their coverage of the unrest and crackdown in Tibet, confusing photos of protests in India and Nepal with protests in Tibet. Rao and his team of young students believed that these errors were not innocent mistakes but instead represented an international media conspiracy against China. The website's first posting, devoted to media errors, generated over 200,000 hits, a long thread of comments, and hundreds of angry young people willing to volunteer their time to help monitor, translate, and analyze what the Western media was reporting about China. When I visited his office in 2009, Rao insisted that the site was not meant to be antiforeign or anti-Western, just anti–Western media, which he believes view China "through colored lenses."

Renamed April Media in 2009, the site has continued to fuel critiques on Chinese social media platforms about Western news coverage of other major events. In February 2011 a member of Rao's online community spotted US ambassador Jon Huntsman on the edge of a crowd near a Beijing McDonald's that just happened to have been designated as a protest spot by people trying to emulate the "jasmine" revolutions of North Africa. April Media wasted no time in disseminating a viral video showing footage of the indignant Chinese netizens confronting the US ambassador. The video ended with a commentary condemning the

United States for interfering in other countries' affairs, and warning that the United States could turn China into yet another Iraq if the Chinese people were not vigilant. The government had no direct hand in this act of "citizen journalism" or the video; it was a spontaneous product of China's nationalist commons.

The motivations of Rao's band of young, nationalistic media activists are similar to those of China's patriotic hackers—often loose alliances of computer whizzes who engage in patriotic hacking missions, and whose relationship to the government is often unclear. In 2009, a team of information warfare researchers affiliated with the University of Toronto uncovered a cyber-espionage network they dubbed "GhostNet," involving at least 1,295 infected computers in 103 countries. They discovered that the Trojan software used to infect these computers was controlled by commercial Internet accounts in China's Hainan province. Targets included the private offices of the Dalai Lama and other exiled Tibetan organizations and offices, from which sensitive and secret information was stolen.

The GhostNet report authors pointed out that it was impossible to prove that the attacks had been organized or carried out by the Chinese government; the best they could do was trace them to a location on China's Hainan Island. Security specialists who study Chinese hacker culture have found that even as government and military cyber-defense and cyber-warfare units have developed and matured, a great deal of freelance hacking continues to be directed at the People's Republic of China's perceived adversaries—with varying degrees of communication (or compensation) taking place between the hackers and Chinese military, state security, or other Communist Party entities. This hacker netherworld works to the government's advantage, providing deniability as well as greater flexibility, in addition to being a ready-made channel for recruitment into real government jobs in the military and security services. In a 2007 book about China's patriotic hacker communities based on several years of in-depth research, security expert Scott Henderson concluded that the bulk of the cyber-attacks taking place up until that time had been conducted by groups of young hackers who had formed into clubs, and of whom the government made both direct and indirect use. "In Chinese society, independence from government direction and control does not carry with it the idea of separation from the state," Henderson wrote. "The PRC government views its citizenry as an integral part of Comprehensive National Power and a vital component to national security." Online communities like Rao Jin's are the "citizen media" equivalent of this phenomenon.

WESTERN FANTASIES VERSUS REALITY

When the United States first began to engage China politically and economically in 1979, many Washington policy makers assumed that capitalism would inevitably bring democratization. Reality has played out differently. Capitalist investors are actually helping the Chinese Communist Party strengthen and refine a gilded cage for China's Internet users. This cage is made as large and exciting as possible for the Chinese people, plying them with ever more digital conveniences, delights, comforts—and even outlets for political debate, giving people the sense that they are much more free than ever before, as long as they do not cross certain lines.

The government cannot afford to sever links between the domestic Chinese Internet and the international Internet without disrupting the international business, trade, and finance upon which its economy now depends. Thanks in part to copious foreign investment, however, the domestic Internet has grown robust and useful enough for the majority of Chinese Internet users that the government has been able to restrict access to the global Internet without widespread public outcry. In early 2011, government network engineers gave the "great firewall" a major upgrade, blocking some of the Internet protocols used to establish secure and encrypted connections and making it much more difficult than ever for people to use virtual private network services to circumvent censorship and evade surveillance. Access to Gmail and a number of other popular international services was not fully blocked but was slowed down to make the sites more difficult to access without delays and connection errors. University and corporate networks suffered disruptions as "punishment" when individual users tried to use circumvention tools to access blocked websites. In May 2011, Hong Kong–based activist Oiwan Lam collected a number of warning notices from corporate and university IT departments and posted them on the Global Voices advocacy site. A typical example was this one issued by the Chinese Academy of Sciences:

> Our faculty's access to overseas websites have been disrupted in the past few days. Upon investigation, the reason is because some users have used circumvention tools to get access to illegal content, hence the public security bureau has black-listed our faculty's IP. Here I remind everyone to follow the rule when using Internet, don't use illegal means to get access to illegal information!

Such measures are an extension of quiet measures implemented several years before on some university campuses. Many of China's top universities, including Peking University and Tsinghua University, provide free broadband to students in their dormitories. But there is a catch: the service is free only for domestic websites and a select set of "white-listed" overseas websites. To access any other websites or services from outside of China, students are charged according to how many megabites they upload or download. Most students, being on tight budgets, see very little of the global Internet.

Fortunately for the government, there are plenty of websites and services on the Chinse Internet to keep people occupied, without ever needing to access sites and services based overseas. Baidu helps them locate all the content on the Chinese-language Internet that their government permits. Social networking platforms RenRen and Kaixinwang substitute for Facebook. People can blog on platforms run by Chinese companies like Sohu and Sina and "tweet" on Weibo. QQ, run by the company Tencent, offers instant-messaging, gaming, and all kinds of interactive services that work seamlessly across both PCs and mobile phones.

Chinese Internet company executives occasionally complain about being forced to censor and track users, but insidiously, their business is too profitable to resist. They are too deeply invested and complicit, as are their international investors. According to the Boston Consulting Group, China's e-commerce market is expected to soon become the world's largest, quadrupling from $71 billion in 2010 to a projected $305 billion in 2015. The amount of money raised by Chinese venture funds—much of it for Internet start-ups—grew from almost $4 billion in 2006 to more than $11 billion in 2010. China's three largest Internet companies, Tencent, Baidu, and Alibaba, are all listed on overseas stock markets. So are several other major Internet brands including RenRen, a social networking platform similar to Facebook, which raised $743 million in its IPO on NASDAQ in early May 2011.

The more successful and popular these companies become, the less incentive Chinese Internet users will have to access blocked foreign services like Facebook, Twitter, or YouTube—beyond a small politically motivated cohort, who are always a small minority of any country's population. It will also become easier to restrict general access to the international Internet without a widespread public outcry. The restriction in turn boosts the traffic and advertising revenue for Chinese Internet companies, making them all the more attractive to foreign investors.

When commercial Internet service first became available in China in 1995, Western policy makers and journalists—myself included—made very different assumptions about how the Internet would change China. In explaining the de-linkage between human rights standards and China's "most-favored nation" trading status in 1994, US Secretary of State Warren Christopher reinforced this assumption when he declared, "Economic development depends upon the free flow of information; computers are the new vehicles of political expression. The software of freedom will, ultimately, prevail over the hardware of repression." Washington policy makers and the US business community failed to anticipate the extent to which the Chinese government could compel domestic and foreign companies to rewrite the software and rewire the hardware to the Chinese Communist Party's specifications.

The Chinese Communist Party has created a system that keeps itself in power while engaging its citizens and helping them succeed economically. The private sector has a stake in maintaining and reinforcing this system with the enthusiastic support of American investors who also profit from its existence. Unless and until a critical mass of Chinese Internet users decide that the status quo is unacceptable and demand access to alternatives beyond their "gilded cage," and unless Chinese entrepreneurs and CEOs decide that it is in their long-term interest to help build a different kind of future, more than likely the Internet's pervasive use in China will actually help *prolong* the Communist Party's rule of China rather than hasten its demise. The implications of this reality extend well beyond China.

CHAPTER 4

Variants and Permutations

When the Egyptian government shut down the Internet on January 27, 2011, it was a serious wake-up call to activists and opposition groups around the world, many of whom had never before fully considered the implications of both government and corporate control over a nation's digital networks. The Egyptian government was able to shut down the Internet relatively easily for two key reasons: First, the government controls the limited number of fiber-optic links going in and out of the country, and the state-owned telecommunications company leases access to them by ISPs. So even if an ISP were to keep its internal links running, the reliance of so many local web and e-mail operations on overseas hosting and domain services means that even most domestic websites cannot load and e-mails cannot be sent within the country when the government cuts off international access. Second, licensing agreements required by the government of all Internet service providers—including foreign companies such as Vodafone—left those companies without legal grounds to challenge a government shutdown order.

Though total Internet shutdown is a drastic measure, it had already been used elsewhere: in Burma during the attempted "saffron revolution" in 2008. In 2009, China shut down the Internet for the entire province of Xinjiang in the country's far Northwest for nearly six months after ethnic riots erupted there. Iran did not shut down the Internet completely during the 2009 postelection Green Movement protests but rather throttled even broadband connections to such slow speeds that it often took hours to upload or download photos and video. Syria shut down large parts of its Internet for a day in June 2011, as protests grew and the government crackdown turned increasingly violent.

The relationship between government and service providers throughout most of the countries in the Middle East and North Africa is very similar to Egypt's, although the corporate service providers—and their loyalties—can vary. After Muammar Gaddafi cut off phone and Internet service to rebel-held areas in eastern Libya in March 2011, a team of engineers working with the rebels found a way to hive off and rewire a part of the government-controlled network so they could operate and control it independently. A major hurdle was that Libya's network was built with equipment from Huawei, a Chinese networking and telecommunications solutions company. Huawei refused to sell them the several million dollars' worth of equipment needed to create an independent network. Ultimately, according to the *Wall Street Journal*, sympathetic governments in the Gulf lent the diplomatic and financial support necessary to obtain the equipment they needed, but in a more circuitous fashion.

The Egyptian and Libyan cases underscore a reality that citizens seeking accountable government anywhere in the world need to understand. Though the technology used for coordinating and organizing may be politically neutral, the context in which it is deployed is rarely so. Governments everywhere—whether they do business in the home government of companies or in the host government of markets—are demanding that Internet and telecommunications companies take sides, or at least stand back and avert their eyes while the government does what it needs to do, leaving the user or customer none the wiser. The notion that companies can focus on delivering value to investors and shareholders and not share responsibility for the broader impact of their business decisions on domestic and geopolitical power struggles—and ultimately the rights, freedoms, and even lives of people around the world—was once quaint but is now obsolete, as standards and expectations have evolved over the past decade.

When it comes to controlling and manipulating the Internet within a nation's borders, most governments in the Middle East and North Africa did not have the same foresight as the Chinese government. Most failed to anticipate or plan ahead—as the Chinese Communist Party began to do as soon as the Internet arrived in China—for the kind of online activism that lubricated and accelerated the protest movements of late 2010 and early 2011. Few other authoritarian regimes can hold a candle to China's lucrative and rapidly growing economy, which is so attractive to international investors. None has been as successful as China in providing fertile ground for a robust ecosystem of domestic Internet and telecommunications companies to take root and grow rapidly, making it so much easier to block foreign web services because the homegrown alternatives are so good at serving the needs of China's businesses and most citizens. Even so, regimes seeking to control and stifle dissent have been learning quickly from one another as well as from China—whose companies sell a growing amount of networking and telecommunications equipment throughout the region—about what can be done.

"CONSTITUTIONAL" TECHNOLOGY

Networked authoritarianism and authoritarian deliberation are not just tools of the Chinese regime; variants can be found worldwide. In a comprehensive book about technology and politics in the Islamic world, the University of Washington's Philip N. Howard argues that "technology design can actually involve political strategy and be part of a nation's 'constitutional moment.'" Howard's research on technology and political change in the Islamic world, conducted over the course of a decade, shows that although the Internet and mobile technologies do not *cause* change, change is unlikely to happen without sufficient mobile and Internet penetration. But there is an important caveat: specifically, *how* governments manage these technologies has a major impact on *whether* activists can actually succeed in using technology to bring about change. The more strategic and forward-thinking a government is in managing and controlling the companies that operate a country's information networks—both in terms of their technical structure and their legal governance—the more likely it is that dissent can be kept within manageable boundaries.

In Iran after the reelection of President Mahmoud Ahmadinejad in June 2009, the notion of a "Twitter revolution" was largely the product of media hype. Even so, social media, including Twitter and Facebook, served as important tools for activists, particularly in bringing global attention to the government's crackdown against the Green Movement when foreign media reporting was heavily restricted. Social media's impact on Iran's internal politics—within which Ahmadinejad is not the most conservative player—is much more complicated. Iran has its own varieties of networked authoritarianism and authoritarian deliberation. Progovernment social media is as active there as in China. As the anonymous editor of the underground e-mail newspaper *Kyaboon* told Global Voices Iran editor Hamid Tehrani in the fall of 2009, social media is just one of the many arenas within which the struggle of "society against an inhumane regime" plays out. But the outcome is by no means certain: "Without struggle," he said, "these technologies can even be of service to the regime."

The Iranian National Guard has gained much ground in that struggle since the Green Movement protests. Even before the elections and protests, research by the OpenNet Initiative, an academic consortium devoted to the study of censorship around the world, found Iran's censorship regime to be second only to China's in its technical sophistication and depth. The regime's control and manipulation of cyberspace prevented the Green Movement from achieving any further political or tactical victories after the street protests of June 2009. As of March 2011, more than five million websites were blocked in Iran. According to the Committee to Protect Journalists, Iran overtook China as the world's top jailer of journalists and bloggers in 2009, then tied with China in 2010. In response to the 2009 protests, the National Guard set up a special cyber-police unit to monitor citizens' online activities, systematizing and centralizing the surveillance and prosecution of "Internet crimes."

Though demonstrators found that text messaging could be a useful organizing tool (except when the regime chose to shut down mobile phone or texting services altogether), they learned the hard way that authorities could track and trace messages. Iranian authorities have also made it clear that activists should assume any digital communications they send—whether by e-mail, voice, or text message—could be intercepted. In 2008, activists released from detention reported having been shown transcripts of text messaging sessions during interrogation. Security officials have made public statements claiming not only to have intercepted unencrypted e-mail exchanges, but also to have successfully decrypted encrypted e-mails. Whether this is true does not actually matter as long as enough people believe it is possible. It was sufficient that a number of people detained in 2009 reported having the contents of their own e-mails as well as other people's e-mails read to them during interrogation. It has also become common practice in Iran to torture people for their Gmail, Yahoo, and Facebook passwords—rendering decryption moot. Spyware and keystroke loggers installed on people's computers also make it possible to monitor even the most heavily encrypted messages. Such reports support the view of researchers and human rights activists that Iran's surveillance program is not only comprehensive but also increasingly effective.

In 2011, the Iranian government announced a move to the next level with the first phase of a "national Internet," also called a "clean Internet" or "*halal* Internet" (*halal* is the Arabic word for "lawful," referring to any object or action permissible under Islamic law). According to communication and information technology minister Reza Taqipour Anvari, a national intranet—a nationwide network that does not connect to the global Internet—would be rolled out in phases beginning in August 2011. It would include domestically managed e-mail service, a national search engine, and other services that Anvari said would enable the government to "better manage national emails and information gathering within the country and to improve security." Another ministry official was quoted, telling Iran's news agency that 60 percent of the nation's homes and businesses would be connected to this internal network by the end of the year and that it would be extended to the entire country within two years. Access to the global Internet could then be restricted to "elites" whose work required it. It remains to be seen whether the plan will succeed. Unlike China, with an abundance of well-financed domestic Internet companies offering alternatives to global services and making it easier to restrict ordinary people's access to the international Internet when needed, large numbers of young Iranians, at least in Tehran, are already addicted to Facebook.

CORPORATE COLLABORATION

Also in 2009, news reports emerged that the technology used to track down activists had been sold to two Iranian mobile phone operators by Nokia Siemens Networks, a joint venture of the Finnish Nokia and the German Siemens. Company executives later confirmed it. The problem, one executive explained at a European Union hearing on the matter, was that the surveillance technology sold to Iran in 2008 is standard "lawful intercept" functionality required by law in Europe, so that police can track criminals. Unfortunately, with the same technology in the hands of a regime that defines "crime" broadly to include political dissent and "blasphemy," the result is an efficient antidissident surveillance machine. Many activists in Iran now report that they leave their phones at home while attending protests or meetings where sensitive political actions could be discussed. The regime's capabilities also mean that the proliferation of mobile video and photos that galvanized global support for the Green Movement protests in 2009 is unlikely to be repeated. People no longer have their camera phones to document the size and scope of protests, or images of police violence against them.

Such a situation is not unique to Iran. A similar "lawful intercept" system sold by Ericsson to Belarusan mobile phone operators is reported to have been put to similar use in the wake of postelection protests in late 2010. The Iranian and Belarusan experiences of 2009 and 2010 undermined previous conventional wisdom about mobile phones and text messaging as the driver of "people power" democratization movements.

Text messaging, first used to bring down President Joseph Estrada in the Philippines in 2001, appeared to be the perfect tool of subversion before governments got smart about anticipating how dissidents would use cell phones to organize political protests. But text messaging can no longer be counted on in countries where the government has the resources and technical capacity to track activity on mobile networks, and where legal and constitutional protections against government interference with networks are weak or nonexistent.

As free speech activist and novelist Cory Doctorow put it in a posting on the popular tech blog BoingBoing, "Mobiles are too closed, the mobile operators too vulnerable to be considered safe enough for use against powerful hostile states. Unless your mobile-driven protest ends with the collapse of the state, it's all too likely that you and your friends will face dire reprisals." The problem with mobiles is that anybody with access to the service provider can easily identify the phone number and unique hardware identifier of each user, then track them in physical space and associate those addresses with an owner's name. This is a major concern even in democratic nations. In nondemocratic ones, the mobile phone is almost useless if a person wants to evade surveillance.

In Egypt, once authorities restored service after shutting down all Internet and mobile phone service for nearly five days, they forced mobile operators—including foreign companies like Vodafone—to send out messages telling people to attend progovernment rallies. It has also become an increasingly common tactic to restrict mobile phone service, or even just specific types of services, such as text and data, in very targeted areas at strategic times. The OpenNet Initiative, citing multiple in-country sources, reported that cell phone text messaging services went down roughly nine hours before Iran's presidential election on June 12, 2009, in an effort to prevent activists from carrying out election monitoring via text messaging. Voice services were down by the next morning as well. Since then, it has become commonplace for mobile service to go down on days and in areas where protests are planned. In fact, as unrest raged across the Middle East and North Africa throughout 2011, reports from bloggers and online activists about the throttling of Internet speeds and periodic shutdowns of mobile text messaging and mobile 3G high-speed Internet functions where those were available have become increasingly routine. In February on the designated "jasmine protest" days in China, bloggers reported that mobile Internet services did not function near the designated protest locations.

The sophistication of technology and techniques authoritarian governments used to hijack citizens' social media and e-mail accounts grew more sophisticated and widespread in 2010 and 2011. Under President Zine El Abidine Ben Ali, researchers considered the Tunisian government's efforts to control online dissent to be among the most aggressive in the world, and in some ways even more so than Iran and China. Not only did the regime employ extensive censorship and surveillance, but government-employed hackers in Tunisia attacked dissident websites, taking them offline, hacked into dissidents' computers and stole information, intercepted and even altered people's e-mails, and hacked into people's webmail and social media accounts. In the months leading up to the fall of Ben Ali's regime, a special program was installed on all Tunisian Internet services to intercept usernames and passwords of everybody logging into Gmail, Yahoo, Twitter, and Facebook. This program enabled government hackers to take over dissidents' accounts, deactivating Facebook pages being used to organize protests and gaining instant lists of everybody they were communicating with.

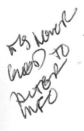

The Tunisian government was able to tamper with e-mails and social media accounts thanks to a technology known as "deep packet inspection." DPI, as people in the industry call it, reads and classifies Internet traffic as it passes through a network, enabling the identification, analysis, blockage, and even alteration of information. In 2008, a group of tech-savvy activists conducted tests that proved the Tunisian government was already using DPI. In 2011, technologists who run the Tor Project, a not-for-profit project that develops software aimed at protecting Internet users' anonymity, concluded in early 2011 that the Iranian government had also deployed DPI technology, based on the Iranian network's ability to selectively block encrypted traffic.

DPI is an invention of companies based in the democratic West and is being deployed by some Internet service providers in many Western countries not only to improve network security, but also to detect illegal file sharing, and even to track user behavior so they can be served with more relevant advertising. Given that the Middle East and Africa are where Internet use is growing most rapidly, many companies that develop and sell these technologies have been marketing their products aggressively in these regions. After the Egyptian government shut down the Internet for several days in early February 2011, Timothy Karr, campaign director of the US-based free speech organization Free Press, uncovered that the Egyptian government had purchased DPI technology from a company called Narus, based in Sunnyvale, California. Narus's most successful product, NarusInsight, which according to the company's website helps "network and security operators obtain real-time situational awareness of the traffic traversing their networks," won a US Department of Homeland Security award in 2010 at the same time that the State Department was quietly urging the Mubarak regime to accelerate democratic reform.

The national telecommunications authorities of Pakistan and Saudi Arabia are just two of the many other authoritarian clients on Narus's global customer list. Narus signed a multimillion-dollar deal in 2005 with Giza Systems of Egypt to license its technology across the Middle East and North Africa to a variety of unsavory governments, including Gaddafi's Libya. Karr's Free Press and the Paris-based Reporters Without Borders also lodged inquiries with Cisco Systems, requesting information about the company's deployment and sale of DPI technologies in countries such as Bahrain, where the company jointly opened an Internet data center with the Bahraini government in 2009, "as an essential component in the drive to improve government services to the populace." Cisco declined to return their phone calls.

Evidence shows that other companies have been actively pursuing business with the region's autocrats. When protesters stormed Egyptian state security offices soon after the revolution, Mostafa Hussein, a psychologist who works with torture victims, discovered a set of confidential documents proving that the Egyptian security forces were viewed as a lucrative client by Western companies seeking to sell surveillance technologies. One such product—which he uploaded in full onto the Internet—was a business offer for an intrusion and surveillance software product called FinFisher, sent by a company called Gamma International UK Ltd. Gamma told the *Washington Times* that the sale did not happen but that the company had broken no laws in trying to sell its surveillance product to Egyptian state security. It is not known which other dictatorships are currently using Gamma's products.

Though China uses its own homegrown programming talent to develop most of the censorship software that makes the "great firewall," many other countries—particularly in the Middle East—have purchased their censorship solutions right off the shelf from American companies. Companies including the California-based Websense, Blue Coat and Palo Alto Networks, Intel's McAfee SmartFilter, and the Canadian Netsweeper all market products that were originally developed to help households and schools shield children from age-inappropriate content or to help businesses block their employees from, for example, visiting Facebook or blogging from work. But these software products are also popular with authoritarian governments seeking to block social, religious, and political content. In a March 2011 report, OpenNet Initiative researchers Helmi Noman and Jillian York disclosed that censorship software developed and sold by North American companies is being used to block social and political content in Bahrain, the United Arab Emirates, Qatar, Oman, Saudi Arabia, Kuwait, Yemen, Sudan, and Tunisia—"effectively blocking a total of over 20 million Internet users from accessing such websites."

Although Websense has a policy that it does not provide governments with filtering tools except in cases where the censorship is restricted to pornography, the OpenNet Initiative still found evidence of Websense's being used by ISPs in Yemen. None of the other companies, however, have shown any public concern for how governments use their products. On the contrary, they appear willing to service their government clients' needs as actively as necessary. In a follow-up story to the OpenNet Initiative report, the *Wall Street Journal* interviewed employees at Internet service providers around the Middle East and North Africa and found that as in China, it is standard practice for government ministries to send them lists of websites and content that they are required to block. Employees then enter the offending website addresses and keywords into the software. But these North American software products are not merely passive tools used to carry out instructions entered by local ISP employees acting on government orders. These software products all come with preestablished censorship lists, in anticipation of clients' needs.

All of these companies maintain lists of tens of millions of websites in dozens of designated categories, from pornography to dating to activism to "alternative news." ISP managers in many countries whose governments have a policy of blocking sexually explicit content tend to automatically switch on censorship for the entire "pornography" and "sexually explicit" categories. Even in North America and Europe, where schools and libraries use filtering software to prevent minors from looking at pornography, free speech and sexual health activists are concerned that sexual education and health websites are also blocked, because of the use of certain words or the display of pictures depicting certain anatomy. But when Internet service providers deploy these commercial software solutions at the national level, the problem is many magnitudes more insidious.

Researchers with the OpenNet Initiative discovered multiple cases in which prominent bloggers who occasionally write about gender issues were mislabeled by SmartFilter and Websense as pornography, causing them to be blocked in many countries around the Middle East. The researchers also discovered that if comments containing hate speech, Viagra advertisements, or links to pornography are posted by anybody on the Internet to a prominent blog, the blog can find itself blocked. OpenNet Initiative researcher Jillian York discovered that her own blog was categorized as pornography by Websense for this reason. The result, York writes, is alarming: off-the-shelf censorship tools like Websense and SmartFilter "can be manipulated by anyone wishing to censor anyone else, just by adding a few links to porn in the comments section."

However, governments in the Middle East and North Africa that have managed to stay in power did not rely exclusively on these Western censorship products to control activists' use of the Internet. A growing number of authoritarian regimes have come to accept that in the Internet age, they have no choice but to allow much more public discourse and deliberation than in the past. Yet, like China, many governments are realizing that their power is not necessarily doomed as a result and have been quick to adopt techniques for manipulating the public discourse.

DIVIDE AND CONQUER

In mid-February 2011 in Bahrain, protesters—mostly members of the Shiite majority—were inspired by events in Tunisia and Egypt to take to the streets. One month later, the Sunni regime unleashed a bloody crackdown with the assistance of troops from Saudi Arabia and the United Arab Emirates. By mid-May more than eight hundred people had been detained, with human rights groups estimating at least thirty people had been killed at the hands of security forces, including four during police detention, one of whom was an award-winning news editor. As a result of the crackdown against online and offline dissenters, Global Voices contributor Ali Abdulemam, who had been arrested previously in 2010, was forced into hiding.

In addition to the violence, arbitrary detentions and arrests, and censorship of websites and blogs run by human rights groups and activists, government supporters launched aggressive efforts to discredit domestic and international critics, deploying tactics that echoed Chinese-style authoritarian deliberation. Not long after the protests began, progovernment bloggers, Facebook activists, and Twitter users popped up like mushrooms after a rainstorm, posting news and "evidence" that the protesters were Shiite terrorists in league with Iran, and blaming them for the bloodshed.

Progovernment activists reposted and retweeted charges by the Bahraini information ministry that opposition media had posted fabricated news intended to provoke sectarian hatred. Mahmood Al-Yousif, a prominent Bahraini blogger and founder of an organization that works to oppose sectarian divisions, was targeted by several blogs and Facebook pages belonging to government loyalists who accused him of inciting hatred. One Facebook page titled "Bahrain Against False Media" listed Al-Yousif as part of a group of people who "attempted to trick the international media and residents and citizens of the country to join them in their hate campaign under the disguise of pro-democracy while in fact it was purely anti-government." Partially as a result of these social media accusations, Al-Yousif was arrested and placed under investigation for several days before being released.

Supporters of the Bahraini royal family insisted loudly that anti-Sunni troublemakers were simply using "human rights" as a ruse to trick the outside world into supporting them in their gambit to overthrow the government and establish a Shiite regime. One Twitter user called @rightsbahrain described himself as a former activist with the Bahrain Center for Human Rights (BCHR), an organization whose members have been persecuted for their efforts to report on human rights abuses. @rightsbahrain said he was "now intent on revealing the hidden agenda about BCHR's political propaganda machine while they use the 'human rights card' as a front." A tweet posted on May 20, 2011, was typical: "I'm pro-truth, pro-democracy (not the way anarchist elements of opposition wants democracy)."

On May 3, 2011, World Press Freedom Day, progovernment social media mavens shared a statement by King Hamad bin Isa Al Khalifa in which he reaffirmed his commitment to "free, impartial and independent Press," stating that "we assure all Press and media figures in Bahrain that their rights are preserved, and that no one will be harmed because of peaceful and civilised expressions of opinion under the law." At the same time, international organizations including Human Rights First, Human Rights Watch, and the International Crisis Group were all reporting widespread violations of human rights and freedom of speech. His words echoed Chinese government statements claiming that the Chinese Internet was as free as anywhere else and was regulated in accordance with international standards.

In Syria, where between March and July 2011 an estimated 1,400 people were killed and at least 15,000 detained in connection with antigovernment protests, the Internet has long been heavily censored. Activist groups had considered the Syrian Internet to be second only to Ben Ali's Tunisia when it came to Internet censorship in the Arab world. Bizarrely, in late February as political tensions mounted, the government suddenly unblocked social media websites such as Facebook, Blogspot, and YouTube for the first time since 2007. The reasons soon became clear: soon after the ban was lifted, government hackers launched what is known technically as a "man in the middle" attack on Syrian Facebook users, inserting a false "security certificate" onto people's browsers when they tried to log into their Facebook accounts through the secure "https" version of the site. This attack enabled government hackers to take over activists' accounts and gain access to their entire network of contacts.

The government also wanted to use social media to get its side of the story across, with the help of citizens claiming loyalty to the government of President Bashar al-Assad. Progovernment Twitter accounts flooded Twitter's #Syria "hash tag"—the place on Twitter where people would most logically look for news about events in Syria—with nonsensical spam. Others directly harassed and lobbed accusations at people expressing support for the protest movement. Though one Facebook page called "Syrian Revolution 2011" quickly gained tens of thousands of followers, other progovernment pages with names like "Youth Only for Assad's Syria" also proliferated.

Even more ominous, the government first tacitly, then openly encouraged progovernment bloggers and hackers to use Facebook, YouTube, and other social media platforms to rally supporters and plan attacks against antigovernment activists, Western media, and governments critical of Assad's crackdown on dissent. In May, an organization called the Syrian Electronic Army (SEA) emerged, describing itself as a group of young Syrians determined to wage war against "the fabrication of facts on the events in Syria." Though the group claimed to be independent, its website was hosted on computer servers belonging to the government-affiliated Syrian Computer Society. The so-called army's members used a series of Facebook pages to recruit new "soldiers" to its cause, targeting the Facebook pages and accounts of government critics and activists with insults and progovernment messages. For the more technically adept recruits, the group distributed denial-of-service-attack software to be used to bring down or deface opposition websites. SEA also went after a range of news, government, and even commercial websites in the United States, Britain, Italy, and Israel, defacing them with messages like "You have been hacked by ArabAttack! Syria Forever." The group's official YouTube page documents the various attacks, accompanied by military symbols and set to patriotic music.

In June, Assad praised SEA directly. "Young people have an important role to play at this stage, because they have proven themselves to be an active power," he said in a speech. "There is the electronic army which has been a real army in virtual reality." The group promptly posted a thank-you note on Twitter, followed by a longer response on Facebook: "Our message to the news agencies and reporters: If you have a shortage of professionals to report the correct news . . . the hordes of the Syrian Electronic Army will not be forgiving with you." Later that week, SEA claimed responsibility for attacking the website of the French embassy in Damascus for the "negative stand of the French government on Syria." They also attacked ten Israeli websites, leaving the message "We Are the Syrian People, We Love our President Bashar Al Assad and we are going to return our Jolan Back, our Missiles will be landing on each one of you if you ever think of attacking our beloved land SYRIA."

Syria under Assad is just one of the many authoritarian regimes that have grown increasingly sophisticated in using the Internet, as much if not more than they are trying to control it. Such tactics bolster the power of ruling regimes by marginalizing liberal or oppositional members of society at the hands of other citizens—either instead of the state or in addition to the state, depending on the country. By emulating and adapting local variants of networked authoritarianism and authoritarian deliberation, governments seek to constrain citizens' ability to build online organizations and communities dedicated to finding viable po-litical or policy alternatives. Democracies—especially new ones in which courts lack genuine independence from the executive branch and op-position parties are weak—are by no means immune to manipulation of the public discourse by the authorities and companies that control digital networks and platforms.

DIGITAL BONAPARTISM

After the fall of the Soviet Union, Chinese government officials enjoyed lecturing foreign journalists and visiting dignitaries about how Gor-bachev had gotten it all wrong: Russia was an economic basket case compared to China, which came to overshadow Russia's influence in the global economy and financial markets. Because Gorbachev had launched glasnost before perestroika, the Chinese argument went, there had been no clear political consensus in the Soviet Union, and then in Russia, about how to reform the economy, and thus Russia struggled to devise a coherent and consistent approach to economic policy. The Chi-nese people, they argued, were much better off than the Russians and others in the former Soviet states, thanks to clear and decisive economic leadership by the Chinese Communist Party. Political reform could come later when the Chinese people had gotten rich and could afford such luxuries.

Ironically, the Internet may be bringing the Chinese and Russian models closer together again. In the *18th Brumaire of Louis Napoleon*, Karl Marx coined the term "bonapartism" to describe political leader-ship by a populist demagogue who seeks to legitimize himself with dem-ocratic rhetoric and trappings. The term was inspired by Napoleon Bonaparte's rule after the French Revolution: usurpation of popular rev-olution by military officers, strong nationalist messages, populist rheto-ric about reform and equality, and elections lacking strong alternatives. Today, after having thrown off communism, Russia is pioneering a new digital-age version of bonapartism: conservative-nationalist rule unchal-lenged by viable political opposition despite the formal existence of multiparty parliamentary and presidential elections, which are effectively populist plebiscites, and manipulation of public opinion through the control—both direct and indirect—of digital networks and platforms.

The Russian government does not filter the Internet as China does: antigovernment websites, blogs mocking Vladimir Putin and Dmitry Medvedev, and social commentaries of the most biting and cynical kind are not blocked from view on the Russian Internet. But there are other ways to control online speech. Controls over offline media—particularly broad antidefamation laws that can be interpreted to include a great deal of critical speech, as well as a statute against extremism—have been extended into the online environment. According to the Committee to Protect Journalists' 2010 *Attacks on the Press* annual report, "Particularly in the Russian provinces, criminal prosecution of Internet journalists is often coupled with other forms of pressure, including intimidation, assault, and murder." Most famously, Anna Politkovskaya, an investigative journalist who had written critically of Russia's policies toward Chechnya and authored a book titled *Putin's Russia: Life in a Failing Democracy*, was assassinated in 2006. In December 2010 in Moscow, journalist and blogger Oleg Kashin was hospitalized after a brutal attack by two unknown persons who had been waiting near his house with a bouquet of flowers. They left him lying with a broken jaw, a broken leg, a fractured skull, lungs full of blood, and fingers torn off at the joints—one of them had to be amputated. The assailants clearly wanted to keep him from writing ever again, and to send a warning to everybody else who might emulate him.

Online and offline writers in Russia are left with good reason to fear for their lives if they push the envelope too far. At the same time, the Russian government does not engage in the direct blocking of websites. Instead, Russian Internet users are constrained and manipulated by a range of what researchers at the OpenNet Initiative call second- and third-generation Internet controls, including surveillance, debilitating cyber-attacks, and proactive manipulation of the public online discourse of the sort that is deployed with growing sophistication in countries like Syria and Bahrain as well as by Iran and China.

These tactics include what human rights groups call "digital violence," committed by people who cannot be directly linked to the government—if they can be identified at all. Hackers frequently launch "distributed denial of service" (DDoS) attacks against websites of government critics and opposition groups, taking them offline at times when the public might be most interested in viewing their content. One example is the use of cyber-attacks to sideline the website of jailed Russian oligarch Mikhail Khodorkovsky, who had challenged Vladimir Putin politically after amassing vast wealth through the oil company Yukos, just hours before the court announced his guilty verdict in December 2010.

Other tactics involve both formal and informal government controls over Russia's telecommunications companies and web businesses. The Russian-language Internet, known popularly as Runet, is serviced mainly by Russian-language platforms, including LiveJournal, which was purchased by the Russian company SUP from the California-based Six Apart in 2007. Yandex, not Google, is the dominant search engine of the Runet. The OpenNet Initiative reports that "informal requests to companies for removal of information" by government officials are common. Warrantless surveillance of Internet users by the FSB (the KGB's successor) is also a built-in feature of the Runet, thanks to a law passed in 2008 requiring Internet service providers to purchase and install equipment that would enable FSB personnel to directly monitor the online activities of specific users. This legalized surveillance capability, according to the OpenNet Initiative, has become "standard" throughout the nations of the former Soviet Union and is believed to be the reason for a number of arrests of Internet users who have posted anonymous antigovernment comments in several former Soviet states.

Even so, the blogosphere has become the lifeblood of Russian political discourse. Internet users are not completely cowed or silenced by any means. With newspapers and television under the government's thumb, the Internet is where Russia's most robust discussions of public policy and the country's political future can take place. The range is broad: there are both nationalist extremist bloggers and progovernment communities. Reformist bloggers expose corruption and malfeasance. In the summer of 2010, a loose network of Russian bloggers created a "Russian Help Map" to coordinate firefighting efforts after it became clear that local fire departments were incapable of handling these efforts on their own. One of Russia's most popular political bloggers, Alexey Navalny, runs an anticorruption whistle-blowing website, Rospil.info.

In Russia, the Internet enables the government to embrace a more populist style—engaging people with a more personal relationship with the government—without actually committing to protect the rights of unpopular dissenters, minorities, and people the regime believes threaten its stability. In 2006 when then-president Putin conducted his first-ever live webcast with Russia's netizens on Yandex, he famously responded to questions about when he had lost his virginity and whether he had ever tried marijuana, in addition to addressing questions about military conscription, immigration controls, and other civic issues. Such interactions are now an accepted component of Russian politics. Current president Medvedev has moved beyond Putin in his embrace of social media, setting up a Twitter account and conducting online chats, as he cultivates a more liberal image in contrast to that of his predecessor.

Yet most activists say they are under no illusions that the Internet and social media are at best driving a slow and circuitous evolution of the political agenda, if not system. As Navalny told Global Voices contributor Gregory Asmolov in an interview in late 2010:

> Actually, Internet for the government is some kind of a focus group. The Russian government is very populist. They just like to do what the people want. I mean, if it doesn't contradict their own interests. The political agenda, however, will be tested on the Internet. And this is why it will have an influence—but no direct impact.

As elections approached, pressure on bloggers like Navalny grew. In late May he was called in for questioning over whether his website's logo desecrated the Russian state emblem. Even more chilling, people who had made donations to his project began to receive mysterious and threatening phone calls. All of them had donated through Yandex.money, an online payment system run by the financial services arm of the search engine. One of these supporters described receiving a call from a woman claiming to be a reporter named Yulia Ivanshova from the newspaper *Our Time*, but who clearly had access to the blogger's account information. The conversation, as translated by Ashley Cleek of Global Voices, proceeded as follows:

> Yulia Ivanshova: Tell me, why do you support Alexey Navalny?
> Me: I don't understand the question.
> YI: Why do you support Alexey Navalny?
> Me: I heard the question, I don't understand it.
> YI: Money was sent from your Yandex account to support Alexey Navalny, is this right?
> Me: (a little taken aback by the lady's knowledge) And where did you get this information from?
> YI: We have our sources, it's public information.
> Me: (even more shocked by the words "public information") Why won't you share with me who these sources are, if it is public information?
> YI: No, I will not. But you didn't answer my question. Why do you support Alexey Navalny?
> Me: Well, then, let's exchange information. You tell me how you know about transactions from my Yandex account, and I will tell why I support Alexey Navalny.
> YI: Who transferred money to you to support Alexey Navalny?
> Me: I took [the money] from my paycheck and transferred it.
> YI: That's not true. Ten minutes before you transferred the money, money appeared in your Yandex account from a Moscow Credit Bank ATM. In the Yandex account from where you transfer money, there is not a single Moscow Credit Bank ATM.

A few days later, the BBC Russian service reported that Yandex .money had handed over the financial and personal records belonging to Navalny's donors to the Russian secret service, or FSB. When queried by journalists, Yandex executives explained that they had no choice in order to remain in compliance with Russian law. One of Navalny's supporters managed to trace the identity of the person who had left him a vaguely threatening message and who had used her real name. He discovered an online profile in which she listed herself as a member of Nashi, the pro-Kremlin youth movement. Navalny and his supporters concluded that the FSB had shared the Yandex.money account information with Nashi, although the group officially denied the allegations. The message to potential supporters of opposition groups is nonetheless clear: Watch out, or you never know who might gain access to your financial transaction records; and who knows what those angry young patriots might do if they decide to take matters into their own hands.

It remains to be seen whether a more genuine form of democracy, in which dissenters' rights are protected from extrajudicial threats and vigilante violence, will emerge from Russia's digital bonapartism. In the meantime, a new model has emerged that can be replicated elsewhere: government leaders use the Internet to carry out a much more direct and populist discourse with citizens in ways that were not possible before the Internet, thus bridging an emotional and psychological gap between rulers and ruled, and building greater public sympathy for the leaders as people. The Internet serves as a "focus group" and early-warning system for the government, alerting policy makers when certain policies just are not working or need modification to prevent unrest. More negatively, progovernment bloggers and journalists are encouraged to mount slur campaigns to discredit reformists and activists who pose serious challenges to the regime's credibility. Anonymous threats against activists who cross the line also help. A full-ranging public discourse about the nation's political future is thus constrained and stunted, and the status quo power arrangements are more easily maintained. This dynamic begs an uncomfortable question: To what extent are the world's oldest, most established democracies vulnerable to digital bonapartism?

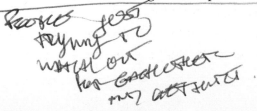

PART THREE

DEMOCRACY'S CHALLENGES

If despotism were to be established amongst the democratic nations of our days, it might assume a different character; it would be more extensive and more mild; it would degrade men without tormenting them.

—ALEXIS DE TOCQUEVILLE,
Democracy in America, 1840

CHAPTER 5

Eroding Accountability

On the day the Mubarak government turned off the Internet in Egypt, Senator Joe Lieberman, chairman of the Senate Committee on Homeland Security, introduced a bill bizarrely named the Cybersecurity and Internet Freedom Act of 2011. An earlier incarnation of the bill, introduced in mid-2010, was much derided by civil liberties and free speech groups as the "kill switch bill." Not only would it grant the president the authority to declare a "cyber emergency," but it also would empower the government to force companies running critical Internet functions to shut down their services. Lieberman told CNN, "Right now, China, the government, can disconnect parts of its Internet in a case of war. We need to have that here, too."

After the Arab Spring of 2011, such statements became politically unfashionable. The bill's revised version emphasized that "neither the president . . . or any officer or employee of the United States Government shall have the authority to shut down the Internet." Lieberman explained that the bill was unrelated to what had happened in Egypt or China but instead would provide "precise and targeted authorities to the president" over critical infrastructure controlled by private companies. Even so, the bill's opponents pointed out that it would give the Department of Homeland Security (DHS) blanket authority to issue orders to private companies in the event of a "cyber emergency," however the president chose to define the term. Companies would be barred by law from challenging DHS orders in court.

As of May 2011, Lieberman's bill was one of roughly fifty Internet-related bills in Congress, with several competing ones on cyber-security alone. Regardless of which prevail or in what form, the deeper problem is that over the past decade, as Americans have grown increasingly dependent on the Internet and cell phones, laws have been passed and policies implemented that make it vastly easier for government agencies to track and access citizens' private digital communications than it is for authorities to search or carry out surveillance of our physical homes, offices, vehicles, and mail. Under two presidential administrations—first George W. Bush and then Barack Obama—standards of oversight, due process, and accountability have been eroded in ways that have made it easier for government agencies to abuse power and more difficult for citizens to hold the abusers accountable.

SURVEILLANCE

The erosion began in earnest when the Patriot Act was passed hastily if understandably in 2001, in the aftermath of the September 11 terrorist attacks. Among other provisions, the Patriot Act allows the FBI to obtain telecommunication, financial, and credit records without a court order.

Another act of legislation, the Foreign Intelligence Surveillance Act (FISA) Amendments Act, granted US companies immunity from being sued by their customers after having complied with blatantly illegal government surveillance of those customers. As a result the American people have had little meaningful recourse against illegal government surveillance of US citizens through corporate networks.

What we do know about the US government's abuse of its power via private networks is largely thanks to whistle-blowers and leakers. One of them was Mark Klein, a technician who retired from his job with AT&T in San Francisco in 2004. Klein worked at AT&T's Folsom Street facility, a major exchange point for phone and e-mail traffic between AT&T's customers and a large number of other networks and service providers around the United States and the world. In early 2003, the US National Security Agency (NSA) built a secret room at that facility. Klein reported that AT&T engineers assisted the NSA in directing all of the facility's e-mail and phone traffic through that room, where it was intercepted and transmitted for further analysis. Another whistle-blower, Justice Department attorney Thomas Tamm, confirmed that similar interception points had been set up to gather and analyze e-mails and phone calls by Americans who were not suspected of any crime. No court order was issued authorizing this surveillance, which, civil liberties advocates point out, makes it illegal. After the *New York Times* first reported Tamm's revelations of illegal, warrantless wiretapping, the Bush administration argued—with its characteristic "you're either with us or against us" logic—that these measures were necessary to fight terrorism in the wake of the September 11 attacks.

Because the FISA Amendments Act grants companies immunity from prosecution for assisting in surveillance, even if it is illegal or unconstitutional, class-action lawsuits against AT&T as a direct collaborator in illegal government abuses of surveillance powers have gone nowhere. The Obama administration surprised many liberal voters by being unsympathetic to these lawsuits: as a candidate, Senator Obama initially pledged to reform the Patriot Act and overturn the FISA Amendments Act, then shortly before his election voted to extend the FISA Amendments Act. As president, Obama also appeared to have changed his mind—or at least his willingness to act—on reforming the Patriot Act. Since then the administration has opposed amendments supported by members of Congress from both parties to tighten judicial and congressional oversight over warrantless surveillance of Americans, and administration lawyers have pushed aggressively to dismiss lawsuits against AT&T and the NSA by civil liberties groups.

US law is very strict regarding phone tapping and physical searches of homes—or even the physical computer hard drives located inside private homes and offices. But it is vague and out-of-date when it comes to digital communications stored "in the cloud": on computer servers operated by service operations ranging from webmail services such as Gmail and Hotmail, to Amazon web hosting, to Blogspot, Facebook, and Twitter. A law called the Electronic Communications Privacy Act (ECPA), which has not been substantially revised since 1986, before the World Wide Web—let alone webmail or social networking—came into existence, allows law enforcement authorities to request all data or e-mail without the need for a warrant if it was stored for more than 180 days. (Back in 1986, nobody imagined that people would want or need to store confidential or private data on centralized servers for longer than that.) Firms often will see it in their interests to collaborate with the government. Such an outdated law makes it more difficult for companies to protect users' rights even when they see a clear business interest in siding with the customer. To address this problem, Google, Facebook, Microsoft, AT&T, and a number of other companies have teamed up with civil liberties groups to lobby Congress to update the ECPA. The goal of the coalition, called Digital Due Process (DDP), is to require a warrant based on probable cause if the government wants to seize contents of a Gmail, Facebook, or Twitter account, enlist the phone company to track the location of your cell phone, or demand IP address information from YouTube.

The DDP scored a victory in May 2011 when the Senate Judiciary Committee introduced a bill that would amend the ECPA, requiring a warrant for law enforcement authorities to access e-mails and other data stored on corporate servers. Locational data about cell phone users, however, would remain fair game. Meanwhile, the Obama administration has been pushing for expanded retention of and easier access to such data. At a May 2011 congressional hearing on mobile data collection and privacy, a Justice Department attorney argued that mobile phone providers should be required to store more detailed data about users' movements, because this would make it easier for law enforcement bodies to catch criminals. Giving police the right to install surveillance devices in the homes of all citizens would no doubt be very helpful in catching criminals too, but there are good reasons that is not permitted without probable cause. It is alarming that the Department of Justice suspends its concern for citizens' privacy and civil liberties when it comes to their digital lives.

Though the US government is required by law to document publicly its wiretapping of phone lines, as of mid-2011 it was not required to do so with Internet communications that are stored on third-party computer servers. Furthermore, until 2009, companies complying with National Security Letters (NSL)—a kind of administrative demand letter requiring no probable cause or judicial oversight—were barred by law from informing customers about the existence of the NSL requests. The constitutional challenge to this gag provision was not mounted by any of the large brand-name Internet and telecommunications companies, which are believed to have received hundreds of thousands of these secret letters over the past decade. Rather, the challenge came from an entrepreneur named Nick Merrill who ran a small, New York–based Internet service company called Calyx Internet Access. In 2004 he received an NSL that he believed was excessively broad and violated his users' rights. He enlisted the help of the ACLU in challenging it. Later that year a federal court ruled the NSL statute unconstitutional as written, prompting Congress to amend the law, allowing recipients to challenge both the demands and the gag orders. Then in 2008 an appeals court ruled parts of the revised law unconstitutional as well, prompting further revisions requiring the government to prove that disclosure of an NSL would harm national security.

In early 2011, Christopher Soghoian, an antisurveillance activist and doctoral candidate at Indiana University, published a research paper analyzing information from as many sources as he could find about government surveillance demands. His conclusion was that "law enforcement agencies now make tens (if not hundreds) of thousands of requests per year for subscriber records, stored communications and location data." He also found that the Department of Justice underreports the volume of requests it makes to companies by "several orders of magnitude." Meanwhile, only a handful of companies have even admitted to the scale of requests they receive.

Research by civil liberties groups raises further questions about whether the FBI can be trusted to use this expanded power responsibly. In January 2011, the Electronic Frontier Foundation (EFF) published a report concluding that, based on its analysis of FBI documents related to investigations from 2001 to 2008, "intelligence investigations have compromised the civil liberties of American citizens far more frequently, and to a greater extent, than was previously assumed." The EFF estimated that based on analysis of documents it obtained through Freedom of Information Act requests, as many as 40,000 violations of law may have occurred during that period. Judicial and congressional oversight of FBI intelligence investigations was found to be "ineffectual." Furthermore, the EFF found that in nearly half of cases in which the FBI abused the use of National Security Letters requesting information, phone companies, Internet service providers, financial institutions, and credit agencies "contributed in some way to the FBI's unauthorized receipt of personal information."

For Americans seeking to change existing laws that they believe are wrong, ill-advised, or unfair, or for anybody who happens to be engaged in dissent, activism, or whistle-blowing that powerful people would like to prevent or contain, such a situation is chilling. Lack of sufficient public oversight and transparency about government surveillance through private networks can dampen dissent and activism over time through the "Panopticon effect," named after a prison designed in 1785 by English philosopher and social theorist Jeremy Bentham. In his design, prisoners are given credible proof that they are under surveillance some of the time, though not all of the time. If they have no way of knowing exactly when they are being watched, they end up having to assume they are under surveillance all the time. In his 1975 book *Discipline and Punish*, philosopher Michel Foucault warned that "panopticism" can extend beyond the physical prison into society at large. When people have unspecific knowledge that surveillance is happening at least some of the time, without clear information about exactly how and when the surveillance is taking place, against whom, and according to what specific criteria, people will choose to avoid trouble and modify their behavior in ways that are often subtle and even subconscious.

In his recent book *One Nation, Under Surveillance*, privacy scholar Simon Chesterman argues that if the liberties of citizens of democracies are to be meaningfully protected, the power of all public and corporate entities with access to citizens' private information must be constrained in a clear and publicly accountable way. "What we are witnessing now," he writes, "is the emergence of a new social contract, in which individuals give the state (and, frequently, many other actors) power over information in exchange for security and the conveniences of living in the modern world." In the Internet age, it is inevitable that corporations and government agencies have access to detailed information about people's lives. Without transparency and accountability in the use of this information, democracy will be eroded.

Meanwhile, societies across the world are at a critical juncture when it comes to governing our digital lives: new institutional and legal frameworks for cyber-security, law enforcement, and civil liberties protections for the digital networks and services citizens depend upon are being debated, decided, and implemented. Standards and norms are being set. In May 2011 the White House released a set of recommendations for national cyber-security legislation that did at least acknowledge the importance of protecting citizens' rights in the course of pursuing other goals. The document recommended that the Department of Homeland Security follow "privacy and civil liberties procedures" in implementing its cyber-security program for protecting private networks against debilitating cyber-attacks that could cripple the national economy. It also stipulated that these procedures be "developed in consultation with privacy and civil liberties experts." Yet it remains unclear whether enough political momentum can be mustered to reverse the erosion of public accountability over the past decade. Of course, this state of affairs has come about because American voters have allowed it—in some cases actively supporting the erosion of civil liberties in the name of fighting terror and crime, or otherwise acquiescing because their attention and priorities have been elsewhere.

It does not have to be this way. In a research paper comparing corporate data retention policies and companies' specific practices in handing over user data to law enforcement and national security agencies, Soghoian documents a wide variance in terms of how much user information is retained, how it is stored, and precisely how it is shared with government agencies. What this means is that companies actually have a great deal of *choice* when it comes to their relationship with government agencies; they can also choose to lobby for laws and regulations that provide greater protection for their customers and users, or not. The problem, Soghoian points out, is that when most people choose their broadband provider, mobile phone service, web-hosting service, social networking service, or personal e-mail provider, company policies and practices in dealing with government surveillance are rarely considered. Part of the reason is that it is very difficult for an ordinary person to know what each company is doing and to compare company practices in a meaningful way.

Soghoian suggests that it is possible to "stimulate a market for effective corporate resistance to government access" by mandating greater transparency on the part of corporations and government. If Congress proves incapable of passing laws to require such disclosure, concerned citizens should press state legislatures to pass such disclosure laws, especially in such key states as California, where many Internet companies are located. The result would be greater protection for all customers of companies based in those states, regardless of the jurisdiction in which customers live.

Even if no legislature passes such disclosure laws, however, greater media scrutiny, customer awareness, and user activism might help to push companies in a more citizen-centric direction. But without clear transparency and accountability about how, when, and under what specific circumstances personal information is being collected and used, citizens have good reason to worry about the growth of the state's "panoptic" power.

WIKILEAKS AND THE FATE OF CONTROVERSIAL SPEECH

WikiLeaks and several news organizations that the whistle-blowing organization had chosen as partners published the first batch of classified US diplomatic cables, allegedly leaked by disgruntled US Army Private Bradley Manning, in November 2010. Vice President Joseph Biden declared WikiLeaks' leader, Julian Assange, to be a "high-tech terrorist." Senator Joe Lieberman declared that "WikiLeaks' illegal, outrageous, and reckless acts have compromised our national security and put lives at risk around the world." Meanwhile, the WikiLeaks "Cablegate" website, dedicated to showcasing the leaked diplomatic cables, came under distributed denial of service attacks of unknown origin. The site was unable to stay online. Assange decided to move the storage and publication of his website's data to another web-hosting service, run by Amazon, which is known for its robustness in defending against cyber-attacks and thus often used by small human rights groups that lack sufficient in-house technical expertise to defend themselves.

Two days after Assange moved the Cablegate site to Amazon, Amazon headquarters received a call of complaint from Lieberman's office. Shortly thereafter, Amazon booted WikiLeaks. The senator responded with a statement: "I wish that Amazon had taken this action earlier based on WikiLeaks' previous publication of classified material. The company's decision to cut off WikiLeaks now is the right decision and should set the standard for other companies WikiLeaks is using to distribute its illegally seized material."

Amazon later insisted that it had acted independently of Lieberman's phone call. At any rate, legally Amazon was off the hook. Controversial speech hosted on Amazon's web servers is not protected in the same way that speech is constitutionally protected in public spaces. The law gives Amazon the right to set its own rules. Amazon's terms of service clearly state that Amazon "reserves the right to refuse service, terminate accounts, remove or edit content in its sole discretion." By clicking "agree," the customer legally consents to "represent and warrant that you own or otherwise control all of the rights to the content" and confirms that the content "will not cause injury to any person or entity."

Despite the absence of criminal charges against—let alone conviction of—anybody involved with publishing the cables, several other companies including PayPal, MasterCard, and EveryDNS proceeded to sever commercial ties with WikiLeaks. Though these companies were within their legal contractual rights to drop any customer for any reason, in dropping WikiLeaks they nonetheless sent a clear signal to their customers: politically controversial speech—even if there is a strong case to be made that it is constitutionally protected—will not always be welcome with us.

It is also important to recognize that some companies tried to resist efforts to silence people who had damaged the US government and its diplomatic effectiveness but who had yet to be formally charged with, let alone convicted of, a crime. In January 2011, the news broke that US government prosecutors had obtained a subpoena requiring Twitter to hand over account information for five people who had been involved with WikiLeaks' publication of classified US diplomatic cables. The subpoena concerned not tweets—which are public to begin with—but private direct messages between users, records of IP addresses (which can help investigators determine the location of the computers or devices from which certain postings were made), and so forth. The government had not established a criminal case against any of these individuals. Twitter's lawyers fought in court for the order to be unsealed, and thus able to be shared without breaking the law. They won, and immediately informed the five individuals of the request made for their information. Asked by the *New York Times* about the case, Twitter spokeswoman Jodi Olson replied, "To help users protect their rights, it's our policy to notify users about law enforcement and governmental requests for their information, unless we are prevented by law from doing so." Several WikiLeaks team members say they suspect that Google and Facebook were served with similar subpoenas. As of August 2011, the two companies had no comment.

The implications of WikiLeaks—and the issues it raises—are extraordinarily complicated, and to be properly understood, they need to be unraveled, with each strand of intention and consequence analyzed separately. Certainly innocent diplomatic sources—including democracy and human rights activists—were harmed or put at risk by the cables' release, and exposing highly sensitive work by capable diplomats may have had a negative effect on relations with governments where the United States has legitimate, even vital foreign policy and national security interests at stake. Yet at the same time, citizens in a number of countries have used some leaked documents to expose government behavior that is clearly unacceptable by any common standards of accountability and responsibility. Regardless of whether one views the intentions and consequences of WikiLeaks' release of diplomatic cables favorably, the US government's response to WikiLeaks highlights a troubling murkiness, opacity, and lack of public accountability in the power relationships between government and Internet-related companies.

In a speech in February 2011, Secretary of State Hillary Clinton sought to distance the State Department—and the US government more generally—from individual politicians and media commentators who called for Julian Assange's head without any apparent interest in due process. She also made the point that the US government did not pressure private companies such as Amazon and PayPal to sever their ties with WikiLeaks. But through subsequent reporting in newspapers and research by civil liberties lawyers, it has also become clear that the companies were influenced by government statements and opinions. These included a letter by State Department legal adviser Harold Koh, in which he wrote that the "violation of the law is ongoing" as long as WikiLeaks continues to publish the leaked diplomatic cables.

As Harvard legal scholar Yochai Benkler pointed out in a group e-mail discussion with colleagues about WikiLeaks and the State Department's actions (which I am quoting with his permission), Koh's assertion was patently "false, as a matter of constitutional law." The Justice Department has not managed to bring a viable case to a court of law against WikiLeaks or any other entity involved with publishing the cables. Benkler argued the government had no case unless it could prove that somebody involved with WikiLeaks directly conspired with Manning.

What Benkler and many other constitutional scholars find insidious about the US government's approach to WikiLeaks is that since the government has no genuine case against the publishers, its assertion of WikiLeaks' illegality—no matter how groundless—"leaves room for various extralegal avenues that can be denied as not under your control to do the suppression work." The Obama administration can deny having directly ordered Amazon, PayPal, EveryDNS, and other businesses to sever ties with WikiLeaks, thus avoiding claims that it has done anything unconstitutional or illegal. But its assertions about WikiLeaks nonetheless succeeded in making it more difficult for a politically controversial organization to publish and raise funds in the United States.

We have a problem: the political discourse in the United States and in many other democracies now depends increasingly on privately owned and operated digital intermediaries. Whether unpopular, controversial, and contested speech has the right to exist on these platforms is left up to unelected corporate executives, who are under no legal obligation to justify their decisions. The response to WikiLeaks' release of classified cables is a troubling example of private companies' unaccountable power over citizens' political speech, and of how government can manipulate that power in informal and thus unaccountable ways. This opaque manipulation is done in ways most people are unaware of or in some cases may even support, because they believe it does not affect them as law-abiding citizens. They may continue to believe that until they or someone they care about find *themselves* to be politically marginalized or vulnerable, or find that their rights have been violated for whatever reason.

CHAPTER 6

Democratic Censorship

Until April 2007, Kathy Sierra wrote a popular blog called "Creating Passionate Users," about how to design software that makes people happy. She was a sought-after speaker at technology conferences. Then death threats and sexually abusive comments drove her to halt all public speaking and writing. In a blog post explaining her decision, she wrote:

> As I type this, I am supposed to be in San Diego, delivering a workshop at the ETech conference. But I'm not. I'm at home, with the doors locked, terrified. For the last four weeks, I've been getting death threat comments on this blog. But that's not what pushed me over the edge. What finally did it was some disturbing threats of violence and sex posted on two other blogs . . . blogs authored and/or owned by a group that includes prominent bloggers.

The threats included altered photos of Sierra with a noose around her neck and a muzzle over her mouth. Some commenters described in graphic sexual language how they would slit her throat and then violate her in horrific ways.

For some reason, anonymous members of a website called meankids.org—who apparently disliked her positive outlook and friendly tone—had decided to target Sierra. Her case is an example of how the Internet can empower malevolent cyber-mobs to victimize innocent people, a disproportionate percentage of them women. The question is:

What can or should government do about this problem? All democracies are struggling to find the right balance between protecting innocent people from bullies and criminals on the one hand and on the other, preserving civil liberties and free expression on the Internet. Unfortunately, many governments are grasping at solutions that put them into conflict with multinational companies as well as human rights groups.

INTENTIONS VERSUS CONSEQUENCES

The Offensive Internet, published in early 2011, offers a collection of essays by prominent American intellectuals who are concerned that the rise of the Internet—and citizens' dependence on it for public discourse—threatens democracy in a number of troubling ways. Unattributed speech, they argue, tends to be irresponsible and inflammatory, causing the public discourse to deteriorate into mudslinging nastiness instead of focusing on issues and facts. Citing cases such as Sierra's, several essays in the book argue that the Internet can make it more difficult for at least some women to participate in public discourse—and hence public life and politics—without being subjected to vicious verbal attacks and even threats. In the essay "Civil Rights in an Information Age," University of Maryland law professor Danielle Citron describes an Internet with two faces: "One propels us forward with exciting opportunities for women and minorities to work, network, and spread their ideas online. The other brings us back to a time when anonymous mobs prevented vulnerable people from participating in society as equals."

Cass Sunstein, writing in his capacity as a Harvard law professor although the book was published while he was serving under Obama as head of the White House Office of Information and Regulatory Affairs, describes how anonymous online speech enables false rumors about public officials and current events to spread like wildfire and become ingrained in the minds of large segments of a nation's or region's population. "The marketplace of ideas," Sunstein writes, "will not work well if social influences ensure that false rumors can spread and become entrenched." He calls for laws to deter people from spreading damaging falsehoods. He concludes that "some kind of chilling effect on false statements of fact is important—not only to protect people against negligence, cruelty, and unjustified damage to their reputations—but also to ensure the proper functioning of democracy itself." Sunstein does not explain how this approach would work, who would have the authority to draw the line between "spreading damaging falsehoods" and publishing controversial analysis or opinions that government authorities believe to be false but that other people genuinely believe to be true, and where such a line would be drawn. Countries best known for punishing people for "spreading rumors" include China, Russia, and Ben Ali's Tunisia.

Sunstein and his coauthors describe problems that are genuinely troubling and that could well erode democracy by reinforcing tyrannies of the majority and driving reason and fact to the margins of the democratic discourse. But the only direct way to prevent inflammatory speech is to eliminate anonymity so that everybody can be held accountable for what they write or upload onto the Internet. As Sunstein must be well aware, American democracy owes its existence in part to anonymity: anonymous pamphlets and tracts like *The Federalist* played an important role in building broader public support not only for a revolution but for a completely untested, experimental, new form of government.

Constitutional lawyer Lee Bollinger, in a recent book advocating a global commitment to protecting free speech, put it this way: "Political majorities and government officials cannot be trusted to exercise the power of censorship in a moderate fashion. Intolerance is natural, especially in times of stress. Given the opportunity to censor, people will censor, particularly when they feel anxious or threatened." Can Sunstein and his coauthors be so naive as to think that power holders in the twenty-first-century United States are different from power holders in any other place or time?

The digital networks and platforms that citizens depend upon are designed, owned, operated, and governed by the private sector. Thus, when democratic governments try to respond to public demands to counter all the "bad" speech online, the job of controlling speech is often delegated to private intermediaries. Yet these private intermediaries are under no obligation to uphold citizens' rights to free expression and assembly. Their interest in guarding anonymous users' identity from government discovery is generally weak, given that the operational costs of defying government orders often appear to outweigh the risks of upsetting some customers—certainly in the short to medium term. If government is empowered to control speech and corporations have little incentive to protect speech, two powerful actors can potentially thwart citizens' freedom of speech in the digital public sphere.

South Korea, one of the most wired nations on earth, with the world's highest high-speed Internet penetration, serves as a cautionary example. In 2005, a young woman was riding the subway with her dog when it defecated on the floor. A fellow passenger proceeded to capture the scene on video as people around her reacted with disgust and outrage when she refused to clean up her dog's mess. The video went viral on the Internet, and she became globally famous as the "dog poop girl." To harass her, cyber-vigilantes quickly discovered who she was and where she lived; she reportedly had to go into hiding, get plastic surgery, and change her identity. Cyber-harassment has already caused a number of celebrity suicides there, according to numerous reports in the national and international media. A national poll in 2006 revealed that 85 percent of South Korean high school students were under stress from cyber-bullying. Not surprisingly, many South Koreans felt that things had gotten out of control, and voters have clamored for their elected representatives to do something.

The result was a law stipulating that all websites with more than 100,000 visitors per day must require users, when creating accounts, to supply not only their real names and addresses but also their national ID card numbers—which happen to be connected to a very efficient national database. Anonymity, South Korean legislators had come to believe, was undermining social stability, enabling cyber-mobs to harass innocent people and cyber-vigilantes to ostracize and shame people for less-than-admirable but nonetheless not criminal behavior. But this legal solution pursued by a democratically elected parliament ended up being used by economically and politically powerful people in South Korea to stifle speech they happened to find threatening.

In early 2009, South Korean blogger Park Dae-sung, aka Minerva, was arrested and jailed for four months on charges of "spreading false information to harm the public interest." His popular and influential postings on one of South Korea's most popular Internet platforms, Daum, provided critical analysis of his country's economic policies and financial situation. Because his writings influenced readers' investment decisions, he was accused by many in the government and media of having undermined South Korea's financial markets in 2008. Park claimed that he merely wanted to write about truths that seemed obvious to him but that the mainstream media were too timid to report—given their close relationships with the regime of President Lee Myung-bak and their need to obtain broadcasting licenses. Herein lies the dangerous slippery slope in legislation to curb anonymity.

Despite the fact that Park published his popular postings under a pseudonym, government investigators were easily able to identify him because he had to register with his real name, address, and ID number. Park was eventually acquitted (the court determined that he believed what he was writing and thus had not intentionally spread rumors), but only after spending four months in jail.

Park's case is only one of many. In early 2010, for example, seventeen people were charged with "spreading false information" after challenging the government's account of how a North Korean submarine had sunk a South Korean warship. This law, plus the real-ID requirement for Internet companies, prompted Google to disable uploading or comments on its Korean YouTube service in 2009. Citing a concern for South Korean Internet users' right to freedom of expression, a statement on the company blog declared, "We believe that it is important for free expression that people have the right to remain anonymous, if they choose."

Though the South Korean Constitutional Court eventually ruled in December 2010 that the law against "spreading false rumors" was unconstitutional, the real-ID registration requirement remained in place until mid-2011. In July the people of South Korea learned a painful lesson about why excessive data retention and ID requirements can make citizens less rather than more secure. The personal information including national ID numbers of some 35 million South Koreans (out of a total population of 50 million) was stolen from the servers of SK Communications, operator of the country's third–most popular Web portal. Security experts traced the attack back to computer servers in mainland China. By early August, the number of South Korean ID numbers available for sale on Chinese websites was reported to have skyrocketed. (Chinese gamers covet accounts on South Korean online gaming sites but cannot gain access without a South Korean ID number, which has created a lively market for South Korean identities.) In the wake of the attack, the government announced that it would gradually phase out the real-name verification policy.

The Indian government's approach to controlling hate speech and suspected terrorist activity online has also raised concerns about whether the costs could ultimately outweigh the benefits. A new law that went into effect in late 2009 holds domestic and international Internet companies—including Yahoo, Facebook, YouTube, and Twitter—accountable for helping to maintain "public order, decency, or morality." Companies are expected to take the initiative to remove potentially inflammatory material. Failure to comply can result in jail terms of up to seven years for executives. The main impetus behind the law is religious violence, an ancient but still current problem in India that can be inflamed by hate-filled postings on the Internet.

Because India's Internet penetration (be it high-speed broadband or low-speed dial-up) remains quite low—under 10 percent of the population—the 2009 law was not a high priority for most Indian human rights groups. But then in April 2011, the Ministry of Communications and Information Technology went several steps further. Under new rules, Internet companies would be expected to remove within thirty-six hours any content regulators designated as "grossly harmful," "harassing," or "ethnically objectionable." Indian free speech advocates have vowed to challenge the rules' constitutionality. As Pranesh Prakash of the Center for Internet and Society in Bangalore put it, "The Indian Constitution limits how much the government can regulate citizens' fundamental right to freedom of speech and expression. Any measure afoul of the constitution is invalid." Google publicly protested the rules in a statement warning that "if Internet platforms are held liable for third party content, it would lead to self-censorship and reduce the free flow of information."

The previous year, Google ran afoul of Italian law as well as public sentiment favoring stronger control of online speech to protect innocent children and the disabled from harassment. In early 2010 an Italian judge handed down criminal sentences to four top Google executives (including David Drummond, the senior vice president who around the same time was busy handling the aftermath of Google's "new approach" to China) because YouTube staff had not been quick enough to remove all copies of a video of an autistic child being bullied by his classmates. The core issue is a tough one for democracies in the Internet age: When awful people put ghastly video on the Internet, with devastating consequences to innocent people, without the consent of the people appearing in the video, who should be held responsible and punished? Google's lawyers argued that staffers acted in good faith and removed the offending video as soon as they were aware of it. The Italian prosecutors countered that Google nonetheless failed to do enough—quickly enough—to protect an innocent child.

Google is not alone; Internet companies around the world face mounting pressure from governments not just to block websites but to delete a wide range of content from the Internet completely, as well as to track what their users are doing so they can be prosecuted or cut off if they do anything illegal. One way to compel censorship and surveillance by companies is to hold them legally responsible for what their users do with their services. The legal term for this practice is "intermediary liability," because the *intermediaries*, or companies transmitting or hosting users' communications or other content, are held *liable* for their users' and customers' behavior. In countries such as China, this arrangement is precisely the legal mechanism that enables an unaccountable government to delegate the bulk of censorship and surveillance to the private sector. In countries with a free press, independent courts, and competitive democratic politics, the problem is less severe.

Even so, civil liberties groups have good reason to be concerned that when excessive liability is placed on intermediaries, companies end up taking on censorship and surveillance functions without sufficient transparency, accountability, or public oversight.

It is at this newly forged digital intersection between corporate and political power where battles over freedom and control are being waged throughout the democratic world. In a January 2011 report titled *The Slide from "Self Regulation" to Corporate Censorship*, the Brussels-based nonprofit European Digital Rights Initiative (EDRI) warned that even though European democracies have not set out to create a "privatized police state," they may inadvertently be heading in that direction, thanks to growing pressure by governments on companies to police themselves. It is increasingly the norm in Europe for Internet companies to have "investigative, monitoring, policing, judging and sanctioning powers delegated to them, occasionally through legislation but, far more frequently, by coercion or by weakening or redefining the protections that they have been able to avail of up until now." As a result, wrote EDRI director Joe McNamee, "intermediaries' own consumers are increasingly being treated as 'the enemy.' Their Internet access is being increasingly blocked, logged, spied upon, restricted and subjected to sanctions imposed by the intermediaries, who fear legal liability for the actions of their clients." The implications, McNamee warns, threaten the core of the democratic enterprise, thanks to "a general abandonment of the traditional concept of the rule of law and the role of the judiciary. The result is the 'death by a thousand cuts' of traditional policing and judicial transparency."

SAVING THE CHILDREN

One example of extrajudicial enforcement is the United Kingdom's Internet Watch Foundation (IWF), a private nonprofit group that collects complaints from the public about websites containing child pornography, then develops a list of banned sites. This list is used "voluntarily" by all of the UK's major Internet service providers, and most of the content on the IWF's blacklist is what most people would consider "legitimately harmful to children." But sometimes the IWF's decisions are controversial: in the fall of 2008 the group's overzealousness made Wikipedia inaccessible for the better part of a day to a large number of British Internet users, because the publicly edited online encyclopedia entry about the rock band Oasis included an album cover depicting an unclothed prepubescent girl. Freedom House, a US-based human rights and democracy organization that tracks global free speech trends, points out that the IWF's "procedures and policies are not transparent. The blocking criteria lack clarity, and the internal appeal process is inadequate. There is no judicial or governmental oversight of the IWF's activities."

Many democracies now deploy national-level filtering systems through which all ISPs (or in some cases most major ones) are compelled to block designated lists of websites to address public concerns about child pornography and other illegal activities conducted on the Internet. The United States does not have a nationwide Internet filtering system, though many school districts, public libraries, and other public networks maintain their own blacklists. But according to the OpenNet Initiative, the number of countries that censor the Internet nationwide has gone from merely a handful a decade ago to almost forty today. This includes the obvious suspects, such as China, Iran, Vietnam, Saudi Arabia, and Tunisia. But the censorship club's fastest-growing membership segment consists of democracies, including the United Kingdom, France, Australia, South Korea, India, and Turkey. Ronald Deibert, director of the Citizen Lab at the University of Toronto, which coordinates much of the OpenNet Initiative's censorship research, wrote in the 2008 book *Access Denied*, "In less than a decade, the Internet in Europe has evolved from a virtually unfettered environment to one in which filtering in most countries, particularly within the European Union, is the norm rather than the exception."

Finland, Sweden, Denmark, and the Netherlands were the first countries in Europe to begin filtering content at the national level (although the Netherlands soon withdrew the policy in response to strong opposition). In March 2011, the French constitutional court upheld a controversial new law giving the Ministry of Interior the power to instruct ISPs to block websites containing child pornography, despite criticisms by free speech groups about the lack of oversight in determining what websites are placed on the blacklist. Then in May 2011, the Law Enforcement Work Party of the Council of the European Union, the EU's central legislative and decision-making body, issued a proposal to create a "single European cyberspace" that would block "illicit content" at Europe's borders.

Concerned that politicians are grasping at ineffectual solutions to a genuine problem, in early 2011 Malcolm Hutty, president of the European Service Providers' Association, wrote a letter to the European Parliament calling for an end to Internet filtering, calling it an "inefficient measure." In debating the issue, some members of the European Parliament pointed out that a website campaigning *against* child pornography was blocked twice in the Netherlands. Complaints abound about "collateral filtering"—the accidental blocking of websites that are unrelated to the stated reason for setting up the filtering system.

Even more disturbing, a growing body of academic research shows that Internet filtering has done little to stop the actual exploitation of real children and may even be exacerbating the problem. In late 2009 a team of academic researchers from France, Germany, the Netherlands, and Ireland published a research paper titled "Internet Blocking: Balancing Cybercrime Responses in Democratic Societies." After examining the impact of censorship on child pornographers, they reached a disturbing conclusion: though Internet filtering makes criminals' websites invisible to the general public, people who are determined to access them can easily figure out how to do so. Furthermore, the censorship does nothing to stop or bring to justice the people who are exploiting children in the first place. Nor does this kind of censorship actually stop criminals from trafficking in children and distributing child porn via e-mail or file-sharing services.

In some countries, concerned citizens have successfully reversed or stopped the implementation of national censorship schemes. In Australia a proposed national censorship system officially aimed at child porn and terrorism met with strong opposition. In March 2009, when the idea was first being debated, WikiLeaks published a secret government list of 2,935 websites that Internet service providers would be required to block as part of a test run. It turned out that the content went beyond child porn and terrorism to include online poker and euthanasia. For reasons nobody could explain, a few businesses with no ties to child porn or other crimes also turned up on the list—including the offices of a dentist in Queensland. The point was that even in democracies, secret censorship lists end up censoring things that go beyond the original mandate—whether by mistake or on purpose. Once websites get on the list, it is difficult for them to be removed, because the list itself is secret.

In 2009 the German Parliament passed an Internet censorship law aimed at protecting children. Free speech groups pointed out that the list of websites to be blocked from public view was maintained by the police without any mechanism for public oversight. Immediately after the law was passed, a number of German politicians suggested that the list be extended to Islamist websites, video game sites, and gambling sites, and book publishers have suggested it would also be nice to block file-sharing sites while they are at it. Once the censorship mechanism was set up, the question became: Could Germany's political and legal systems prevent this mechanism from being abused? Civil liberties groups argued that abuse was inevitable. They eventually won over enough members of Parliament, and the law was overturned in early 2011.

It is thus an undeniable fact that democratic societies face urgent problems—sometimes magnified or accelerated by the Internet—for which voters are demanding solutions from their leaders. Politicians tend to grasp at solutions like censorship and surveillance because they seem expedient and practical in the democratic context but nevertheless invite abuse. Yet solutions to these problems must not make it more difficult for dissent and protest by weaving censorship and surveillance deeply into the legal and technical fabric of the global Internet. It is imperative that voters, politicians, and companies of the world's democracies gain greater awareness of the need to find innovative ways of addressing problems that will not require citizens to pay for security with their freedom.

CHAPTER 7

Copywars

In March 2010 I was invited to address Chinese Internet censorship and the threats faced by digital activists around the world at a hearing organized by the House Foreign Affairs Committee. Aspiring to be all things to all constituents, the hearing was ambitiously titled "The Google Predicament: Transforming US Cyberspace Policy to Advance Democracy, Security, and Trade."

Democratic Congressman Howard Berman, whose district includes parts of Hollywood, chaired the hearing. Not surprisingly, he is a strong advocate for the entertainment industry's interests in Congress, particularly their fight against the copying and sharing of pirated movies and music through the Internet. Next to me at the witness table sat Google Vice President and Deputy General Counsel Nicole Wong. Two years earlier, the *New York Times* had nicknamed her "the decider." Every day Google receives hundreds of requests from governments, companies, and individuals all over the world to remove content. Most situations are handled by staff according to clear-cut procedures for dealing with intellectual property complaints, violations of community guidelines against violence and hate speech, and laws defining illegal speech in countries where Google offers localized services. But when political issues arise—particularly when requests to take down or block material come from governments—Wong is called in to decide what to do.

After listening to Wong's account of Google's decision to stop censoring its search engine in China, several members of Congress praised

Google for its principled stand. Others, however, harangued Google for failing to adequately protect the intellectual property of American entertainment and media companies. Republican John Boozman of Arkansas chastised Wong for not doing enough to stop counterfeiters from using the Google AdWords advertising system. Representative Mike McMahon, a Democrat from New York, cited other testimony that 82 percent of all software used in China is pirated—all obtained, he charged, "through the Internet of course and through using the Google platform." McMahon demanded that Google do more to "protect that vital American interest."

A seasoned lawyer, Wong did not change her expression, but sitting next to her I could sense a slight stiffening of posture. This turn in the line of questioning seemed to have caught her by surprise. A colleague sitting behind her handed her a note with one word scribbled on it: "Italy." Wong then described how three Google executives had just been convicted in Italy because YouTube had failed to act quickly enough to prevent the uploading of a video depicting an autistic child being bullied. If the US Congress were to hold Google's platforms similarly liable for everything their users uploaded or transmitted, Wong argued, the value of Google's services would be severely diminished. "If we had to pre-screen, then our platforms couldn't exist at the level of robustness that enables protesters to upload video from Iran and Burma." As a result, activists' ability to use social media to push for human rights and democracy "would be dampened."

As representatives continued to pummel Wong for Google's failure to adequately police users who violate copyright, she repeatedly emphasized the free expression trade-off. "There is a significant legal infrastructure for protecting intellectual property which we think is appropriate . . . but we also believe that there has to be a balance. Part of our reason for being here at this meeting is to talk about the lack of a similar infrastructure for platforms for free expression, because this has been an area where we believe that in the past the legislation has not paid enough attention. We do believe in the protection of intellectual property. We also believe in the balance that permits a free and vibrant platform for free expression."

Wong's arguments were met with dubious frowns on the faces of the elected representatives of the American people. The cognitive dissonance on display at that hearing highlighted an inconvenient reality: politicians throughout the democratic world are pushing for stronger censorship and surveillance by Internet companies to stop the theft of intellectual property. They are doing so in response to aggressive lobbying by powerful corporate constituents without adequate consideration of the consequences for civil liberties, and for democracy more broadly.

SHUNNING DUE PROCESS

In a January 2010 op-ed in the *New York Times*, Bono—the rock star of all rock stars who has devoted countless hours and large sums of money to combat global poverty and HIV/AIDS—called on Western governments to save the Internet from pirates. "We know from America's noble effort to stop child pornography, not to mention China's ignoble effort to suppress online dissent," he wrote, "that it's perfectly possible to track content."

Statements like Bono's may not intend to be a gift to the Chinese Communist Party, but they are. The reality is that such commentaries expressing implied support for surveillance and censorship—even if for the purpose of protecting intellectual property, not for silencing dissidents— make it easier for Chinese officials to say such things as "the Chinese government's legal management of the Internet is in line with international practice." That is exactly what Foreign Ministry spokesperson Jiang Yu said at a press conference in mid-2011.

To be clear, my point is not to argue against reasonable intellectual property protection. The problem, as with cyber-security and child protection, is one of balance. Companies, artists, and writers (myself included) understandably want to protect their intellectual property. It is harder to understand why the need to protect intellectual property has become a higher priority for many elected lawmakers than due process— the presumption that a person is "innocent until proven guilty"—without which it is very difficult to prevent government authorities from abusing their power against citizens who do things that powerful people dislike but that are arguably legal.

For example, in May 2011, the US Senate Judiciary Committee introduced a bill called the Preventing Real Online Threats to Economic Creativity and Theft of Intellectual Property Act of 2011, otherwise known by its acronym, PROTECT IP. The bill empowered the government and copyright owners to seek injunctions against websites that they believe are "dedicated to infringing activities." They would also be empowered to obtain court orders against intermediary companies providing services (such as hosting, domain name registration, or payment processing) to alleged infringers. Accusers could further compel search engine companies to remove all links to allegedly infringing websites.

Though Judiciary Committee Chairman Patrick Leahy claimed that the goal is to "crack down on rogue Web sites dedicated to the sale of infringing or counterfeit goods," engineers involved with the Internet's creation worried that some of the bill's technical solutions requiring ISPs to reroute traffic away from allegedly infringing websites could degrade "the Internet's value as a single, unified, global communications network." Civil liberties and free speech groups argued that the bill lacked safeguards to prevent abuse of copyright claims in ways that could chill political speech or stifle other speech that is arguably legal. Upon reviewing news coverage of the issue over several months, the *New York Times* editorial board declared its opposition to the PROTECT IP bill as it stood in mid-2011, concluding, "If protecting intellectual property is important, so is protecting the Internet from overzealous enforcement."

Events leading up to the bill's introduction provided plenty of reason to doubt that enforcers can keep their zeal in check. In November 2010 the US Department of Homeland Security's Immigration and Customs Enforcement unit (known as ICE) shut down eighty-two websites without warning, on suspicion of copyright infringement. The *New York Times* reported that one of the websites shut down because ICE considered it to have been used "to commit or facilitate criminal copyright infringement" was a popular hip-hop blog, dajaz1.com. The website owner, known publicly online as Splash, said he was given no opportunity to appeal the shutdown before it happened. The affidavit claimed that the action had been taken because the site provided music without authorization by the copyright holders. Yet Splash showed *New York Times* reporter Ben Sisario copies of e-mails in which employees of record labels and third-party marketers offered him songs mentioned in the affidavit.

Another website shut down for infringement, onsmash.com, appeared to have been targeted on even shakier grounds. A website posting cited in the affidavit provided a link to music by rapper Kid Cudi, with the line "You can pre-order the album on iTunes tomorrow and receive a bonus track on the day of release." Yet another targeted website was Torrent-finder.com, a search engine for file-sharing websites. It does not actually host any of the infringing content itself, and not all content on file-sharing sites is infringing (and though much of it is, it is also important to note that people in the Middle East and Asia also use file-sharing sites to distribute banned political and religious content). The site's Egyptian creator pointed out that people can also use Google and Yahoo to search illegal file-sharing websites, yet they have not been taken down. In raising this case I do not wish to debate the finer points of Torrent-finder's guilt or innocence or its real intent to commit criminal activity. Although proponents and supporters of ICE domain seizures argue that these measures only pertain to "the worst of the worst" actors who are "clearly infringing," the reality is that a great deal of the material on the Internet defined in such a way by law enforcement and copyright holders can in fact be subject to reasonable dispute. Therefore, due process remains essential and must be defended robustly if erosion of free expression online is to be avoided.

The need for robust due process is all the more important given the numerous cases in which copyright infringement claims are used—sometimes blatantly—to chill political activism and social commentary. Some of the most outrageous incidents in the United States have been collected by the EFF on the website Takedown Hall of Shame. In one of these cases in 2009, the whistle-blower website Cryptome published a leaked copy of Yahoo's Compliance Guide for Law Enforcement, detailing prices and procedures for law enforcement and spy agencies requesting information about Yahoo subscribers. Yahoo lawyers wasted no time in issuing a takedown demand under the Digital Millennium Copyright Act (DMCA), which grants Internet companies immunity from prosecution but only if they comply with the takedown request upon receipt. The takedown demand can be challenged in court, but this can be time-consuming and costly for the content-hosting company. Also, in cases where the content in question is politically time sensitive, temporary takedown of a few weeks or months is sufficient to suppress the political discourse.

In another case, Diebold Election Systems, maker of electronic voting machines, sent DMCA cease-and-desist letters to dozens of ISPs hosting websites that had published leaked documents containing internal memos between Diebold executives. All of the ISPs complied at once without a fight except one, the nonprofit Online Policy Group (OPG), which also happened to be hosting a website affiliated with the independent citizen journalism organization Indymedia. OPG sued with EFF's help and eventually won. As OPG Executive Director Will Doherty observed, "Diebold's claim of copyright infringement from linking to information posted elsewhere on the Web is ridiculous, and even more silly is the claim that we as an ISP could be liable for our client's web links." Despite OPG's victory, however, these types of cases have persisted and even expanded into social networking services such as Facebook, where the technology magazine *Ars Technica* found its popular Facebook page shut down in May 2011 as the result of a DMCA complaint whose validity the website's administrators failed to investigate before disabling the page.

AIDING AUTHORITARIANISM

Anticensorship groups who work with dissidents in authoritarian countries—where governments include political dissent in their definition of "illegal" and "infringing"—point out that when the United

States and other democracies demonstrate lack of concern for protecting free speech in pushing for stronger copyright enforcement, authoritarian governments are empowered in their efforts to curb any content they want to categorize as "infringing." Especially in China, strong pressure from the US Trade Representative and US business groups to crack down on copyright violation has had the unfortunate—if unintended—consequence of complementing the Chinese government's efforts to stifle dissent.

After conducting interviews with Chinese officials charged with enforcing copyright, analyzing Chinese policy and legal documents, and observing government antipiracy campaigns for an in-depth study on the politics of Chinese copyright law, Cornell University political scientist Andrew Mertha discovered that Chinese authorities routinely combine intellectual property enforcement campaigns with broader efforts to stamp out not only pornography but also antigovernment material. This approach is easily justified because all creative works banned by Chinese authorities are "by definition an infringement of copyright."

Since the early 1980s, when China first opened up its economy to international trade and investment and found itself needing to create an intellectual-property legal regime from scratch, Chinese legal scholars argued that such a legal regime would foster creativity and innovation needed in a market economy and satisfy the demands of Western trade partners for intellectual property enforcement. It would also be useful for controlling China's political discourse. After the 1989 Tiananmen Square crackdown, close analysis of Chinese official documents revealed that authorities viewed American pressure on China to tighten intellectual property enforcement and improve copyright law as politically convenient. Lack of adequate control over books, periodicals, and other forms of media was considered a major driver of the 1989 political reform movement. Antipiracy campaigns provide excellent cover for stamping out everything the government chooses to define as illegal.

Now in the Internet age, efforts to stamp out online piracy continue to overlap with political crackdowns. Interpretation of copyright law is highly political. For example, the concept of fair use, which allows people to use excerpts of copyrighted works for parody or political commentary, is applied selectively by Chinese authorities. In 2008, the Chinese government invoked intellectual property law as an excuse to stamp out online parodies and critiques of the Beijing Olympics.

That said, it is certainly a fact that rampant piracy of software, movies, and music is epidemic in China. My point is not that the Chinese government should do nothing to protect intellectual property. The point is that US trade policies pushing for tougher enforcement have been blind to the irony that although some parts of the US government—particularly the State Department—are highly critical of Chinese censorship practices, US trade and commercial interests are simultaneously pushing for enforcement practices that actually give the Chinese government justification to maintain those same censorship systems and practices. Although there are no easy solutions to this conundrum, doing nothing about it certainly sends a cynical signal through deeds if not words about America's priorities.

Globally, the US government continues to negotiate trade agreements that will make it easier for governments around the world to punish people for uploading and downloading content deemed illegal. Chief among these efforts has been the Anti-Counterfeiting Trade Agreement (ACTA), which the US government spent four years negotiating in secret with thirty-four other countries. ACTA's Internet section—drafted by the United States at the strong behest of the entertainment lobby— sought to require Internet intermediaries to police user-generated content, cut off Internet access for copyright violators (without due process in a court of law), and remove content that allegedly violates copyright without requiring proof of violation. The text was, bizarrely, declared classified by the US government and kept out of the press until May 2008, when WikiLeaks obtained a leaked copy and published it online. A loud outcry from free speech groups ensued.

Public pressure, particularly among free expression groups in Europe, eventually led to the watering-down of this section, and the most controversial language was removed. But one of the many unfortunate consequences of the entire ACTA process was the reinforcement of the impression among digital free speech activists around the world that when trade negotiators are faced with a choice between copyright enforcement and protecting free speech, economic concerns trump human rights. A research report by the Social Science Research Council examining the impact of US intellectual property enforcement trade policies in Russia, Brazil, India, South Africa, Bolivia, and Mexico documented how the US entertainment and software industries are pushing, with US government support, for "expanded police powers and the wider application of criminal law" in these countries. "The private direction of public enforcement is also problematic on a number of levels," wrote author Joe Karaganis, "and raises concerns about accountability, fairness, and due process."

In Russia specifically, intellectual property enforcement has been used blatantly as an excuse to crack down on dissent. In September 2010, the *New York Times* broke the story that Russian security services had carried out dozens of raids against outspoken activist groups and opposition news organizations, confiscating computers under the pretext that they were using pirated Microsoft software. The *Times* cited Russian court documents showing direct support for such operations by lawyers employed by Microsoft, which had indeed been pushing Russian authorities to do more to protect its intellectual property. Upon being presented with evidence that these enforcement measures were being abused for political purposes, Microsoft headquarters announced sweeping changes in its corporate policies and practices to guard against such abuse. Senior Vice President and Corporate Counsel Brad Smith wrote on the company blog, "We want to be clear that we unequivocally abhor any attempt to leverage intellectual property rights to stifle political advocacy or pursue improper personal gain." Ensuring that free speech is not suppressed under the guise of intellectual property enforcement around the world, however, will require much broader and concerted effort and genuine concern for free speech and human rights by the entire software and entertainment industries. To date, there is little sign of any such concern.

LOBBYNOMICS

In France, President Nicolas Sarkozy, married to the pop singer and model Carla Bruni, a long-standing advocate of strong copyright enforcement, has chivalrously taken it upon himself to defend his wife's interests and lead the European charge against Internet piracy. Sarkozy's government has created a new agency, HADOPI, to supervise ISPs that are now required to monitor their customers for piracy and to send warnings to violators. If the violator ignores all warnings and persists, his or her Internet service will be cut off. A key component of this system is surveillance: mechanisms used by ISPs to determine whether an Internet user is downloading or uploading pirated videos and music.

HADOPI is under aggressive challenge by French civil liberties groups that are concerned that there are few safeguards or public oversight mechanisms to prevent this surveillance capability from being abused for either commercial or political purposes. Those concerns were heightened in June 2011, when the company, which was contracted to serve as a clearinghouse for user information that other companies collected to track infringement, was hacked, causing all the information to be publicly exposed. Such incidents have prompted further concern that systems being put in place to track copyright violation will actually make citizens more vulnerable to more serious crimes, such as identity theft.

Despite these problems, similar schemes are being considered or enacted around Europe. The Digital Economy Act, passed by the UK Parliament in April 2010, involves surveillance and tracking features similar to France's HADOPI. In early 2011 two British ISPs, BT and Talk-Talk, challenged the legislation, arguing that it would potentially infringe on "the basic rights and freedoms" of millions of their customers. Broadband analyst Mark Jackson warned that enforcing the act's provisions could "lead to the blocking of legitimate websites, service speed restrictions, limits on open Wi-Fi usage, account disconnection from your internet service provider or disclosure of private personal details to rights holders for legal action." The challenge was denied in May 2011, although efforts by free speech groups to repeal the law continue.

Also in May 2011, an independent review commissioned by Prime Minister David Cameron found the UK's intellectual property laws to be outdated and dysfunctional. The report, written by University of Cardiff professor Ian Hargreaves, warned that the existing system is holding back innovation, particularly by small businesses and start-ups. Hargreaves found that laws and policies were being formulated not on the basis of solid research and objectively obtained data, but through a process of what he called "lobbynomics." As the entertainment and software industries continued to push for tougher intellectual property enforcement as a matter of greatest urgency for the future of the British economy, Hargreaves found, the evidence was less conclusive:

> No one doubts that a great deal of copyright piracy is taking place, but reliable data about scale and trends is surprisingly scarce. Estimates of the scale of illegal digital downloads in the UK ranges between 13 per cent and 65 per cent in two studies published last year. A detailed survey of UK and international data finds that very little of it is supported by transparent research criteria. Meanwhile sales and profitability levels in most creative business sectors appear to be holding up reasonably well. We conclude that many creative businesses are experiencing turbulence from digital copyright infringement, but that at the level of the whole economy, measurable impacts are not as stark as is sometimes suggested.

Hargreaves called on the British government "to ensure that in future, policy on Intellectual Property issues is constructed on the basis of evidence, rather than weight of lobbying, and to ensure that the institutions upon which we depend to deliver intellectual property policy have clear mandates and adaptive capability."

Lobbynomics, particularly in the case of intellectual property law, is epidemic throughout the democratic world. The United States appears to have a particularly bad case. According to public records, the recorded music industry, led by the Recording Industry Association of America, spent a combined $17.5 million in congressional lobbying in 2009 alone.

Harvard professor Lawrence Lessig points out that members of Congress spend 30 percent to 70 percent of their time raising money and that roughly 50 percent of US senators eventually become lobbyists themselves, perpetuating the cycle.

Of course, the interests of the people who supply the bulk of congressional campaign funds are not the same as those of the people as a whole. Research by Princeton political scientist Martin Gilens points to evidence that Americans—and by extension everybody around the world who is affected by American copyright and trade policies—have good reason to be concerned. After analyzing nearly 2,000 questions about proposed policy changes asked in nationwide polls between 1981 and 2002, and then comparing those answers to actual policy outcomes, he discovered that "when Americans with different income levels differ in their policy preferences, actual policy outcomes strongly reflect the preferences of the most affluent but bear virtually no relationship to the preferences of poor or middle-income Americans." Such findings inspired Lessig, who had been working for the past decade on projects related to copyright reform, to change tack and launch a new organization dedicated to reforming the way in which political campaigns are funded. Until this fundamental flaw in American democracy is addressed, Lessig concluded, convincing lawmakers to pass more evidence-based, balanced, and citizen-centric laws governing people's digital lives will continue to be an uphill—if not futile—struggle.

The rise of the digital commons—without which there would be no Internet or World Wide Web as we know them today—poses a fundamental challenge to the entire system of copyright law, which was conceived at a time when it was difficult and expensive to duplicate and distribute creative works. But if democratically elected leaders adopt the policies pushed by lobbyists and make Internet and telecommunications companies vet and track their users to stop all "infringing" activity, not only can dictators breathe a sigh of relief, but so can incumbent politicians everywhere who would rather not have to face Internet-organized grassroots citizens' movements.

In late 2010 Tim Berners-Lee, inventor of the World Wide Web, suggested that it is time for humanity to upgrade the principles first articulated eight hundred years ago in the Magna Carta: "No person or organization shall be deprived of the ability to connect to others without due process of law and the presumption of innocence," he wrote in an impassioned essay. It is a moral imperative for democracies to find new and innovative ways to protect copyright in the Internet age without stifling the ability of citizens around the world to exercise their right to freedom of speech, access information they need to make intelligent voting decisions, and use the Internet and mobile technologies to organize for political change. Balanced, citizen-centric solutions will require innovation, creativity, and compromise. Sadly, the elected leaders of the world's oldest democracies are disappointing the people who could most use their help by demonstrating very little enlightened leadership and a great deal of short-term self-interest.

PART FOUR

SOVEREIGNS OF CYBERSPACE

Of course, money cannot buy freedom,
but freedom can be sold for money.

—LU XUN, 1923

Don't say that he's hypocritical,
Say rather that he's apolitical.
"Once the rockets are up, who cares
where they come down?
That's not my department," says Wernher
von Braun.

—TOM LEHRER, from the song
"Wernher Von Braun," 1965

CHAPTER 8

Corporate Censorship

In the fall of 2009, Apple officially launched the iPhone in China in partnership with a domestic mobile carrier, China Unicom. As a condition for entry into the Chinese market, Apple had to agree to the Chinese government's censorship criteria in vetting the content of all iPhone applications—or "apps"—available for download on devices sold in mainland China. (Most apps are created by independent developers—individuals, companies, or organizations—then submitted to Apple for approval and inclusion in its app store.) On Apple's special store for the Chinese market, apps related to the Dalai Lama are censored, as is one containing information about the exiled Uighur dissident leader Rebiya Kadeer. Apple similarly censors apps for iPads sold in China. So much for that revolutionary, Big Brother–destroying 1984 Super Bowl ad. Fifteen years later, Apple seems quite willing to accommodate Big Brother's demands for the sake of market access.

Internet and mobile telecommunications companies (whose functions and relationships are increasingly intertwined) create computer code that functions as a kind of law, in that it shapes what people can do and sometimes directly censors what they can see. Sometimes the computer code is written to comply with government demands, as in the Chinese case. Sometimes government uses legislative code—law—to force the evolution of computer code in one direction or the other. Sometimes these laws reflect the "consent of the governed," and sometimes they do not, depending on what sort of government is doing the lawmaking. Sometimes the laws are arguably more reflective of certain powerful and wealthy lobbies than the public as a whole. Sometimes the law intends to reflect the wider public interest but actually reflects politicians' ignorance of technology, thus creating unintended and sometimes negative consequences. At any rate, legislative code does not cover most aspects of computer code, and it certainly fails to anticipate innovations, as Lawrence Lessig first pointed out a decade ago. Neither citizens nor businesses have an interest in excessive micromanagement of computer code by legislative code—or innovation would be impossible. Thus, much of the time companies apply their own criteria, based mainly on commercial factors. Unfortunately, the impact of decisions on customers' ability to exercise their political rights and freedoms is at best an afterthought, if the question even comes to managers' and engineers' minds at all.

NET NEUTRALITY

In 2007, the abortion rights group NARAL Pro-Choice America sent out text messages to supporters who had signed up to receive them. One message said, "End Bush's global gag rule against birth control for world's poorest women! Call Congress. (202) XXX-XXXX. Naral Text4Choice." Verizon Wireless blocked the NARAL messages.

When challenged, company executives said it was company policy to block messages from any group "that seeks to promote an agenda or distribute content that, in [Verizon's] discretion, may be seen as controversial or unsavory." Then, after further news coverage and public outcry, a spokesperson said that the initial explanation had been "an incorrect interpretation of a dusty internal policy . . . designed to ward against communications such as anonymous hate messaging and adult materials sent to children." Verizon did not actually need to explain itself, however: standard customer agreements generally include language stating that the customer understands that service can be cut off and data transmissions blocked for any reason.

In late 2010, T-Mobile blocked a medical marijuana dispensary listing service based in California (where medical marijuana is legal) from using its short code text messaging service, claiming that messages had been sent on behalf of a group of which T-Mobile did not approve. Though the group sued T-Mobile, the company called the lawsuit "without merit." (The case was dismissed by the Southern District Court "with prejudice" against the defendant, T-Mobile.) Indeed, although US law prohibits phone companies from blocking calls placed on their networks, the law does not apply to data and text messages, so as far as the law is concerned, companies are free to do as they please.

Free speech groups point to other troubling examples of political censorship by companies: In 2007 during a live-streaming broadcast of a Pearl Jam concert, AT&T muted the sound when lead singer Eddie Vedder made comments critical of then-president George W. Bush. Comcast has been accused of blocking entire peer-to-peer applications because some people were using them to share copyrighted works illegally, thus hindering the legal use of these networks by people trying to share large video files, which inevitably include homemade videos containing political speech. Granted, in the United States people have other ways of expressing themselves and getting information out. But the fact that citizen efforts to conduct legal political activities over private networks can be arbitrarily stymied by private service operators is an example of how the political discourse in a democracy can be eroded if citizens are not vigilant about exposing political discrimination when it occurs and holding companies publicly accountable.

There is a further concern that even when companies do not discriminate against specific content, they can still discriminate for and against different categories of Internet traffic. This ability could potentially make it more difficult for independent and nonprofit citizen media to reach large audiences or build broad communities. If ISPs are allowed to discriminate, they can also block other content services that happen to compete with their own.

Internet service providers now have the technical ability to "see" what kinds of application a particular subscriber is using at a given moment in time: Are you using Skype? YouTube? Streaming movies on Netflix? Or are you accessing a nonprofit website that uses the open-source

WordPress platform? Your service provider knows. Currently in most countries, there is no law preventing Internet service providers from discriminating between services for a profit: your ISP could in theory offer a "tiered" access package in which access to certain services belonging to major brand names (Amazon, Netflix, YouTube, and Facebook, for example) would be cheaper than access to the general Internet or to lesser known applications. Brand-name companies could pay the ISP for the privilege of gaining preferential access to users at lower or no cost. Or an ISP could demand money from popular services in exchange for smooth and fast delivery.

In late 2010 Comcast attempted to move in this direction when it demanded that Level 3 Communications, a major client of Netflix (which also competes with Comcast's own video services, particularly now that Comcast has acquired a controlling stake in NBC Universal), should pay an extra fee in exchange for smooth delivery of Netflix videos to Comcast broadband subscribers. Level 3 accused Comcast of trying to set up what amounts to a toll booth on the Internet. The Internet's empowering nature has been based largely on the fact that any citizen can potentially create media and reach global audiences just by sending a tweet or publishing a blog. On a "neutral" Internet, it doesn't matter what platform the citizen uses—that person's video, blog post, or podcast is equally accessible to anybody anywhere (as long as a government isn't trying to censor it). On a nonneutral Internet in which platforms and applications must compete by paying fees to gain maximum access to Internet users—and in which Internet users are asked to pay different amounts for different tiers of access determined by the different deals the ISP has cut with the companies that run different services and platforms—citizens will depend all the more on large, well-financed, commercially operated platforms to carry out political discourse and political organizing.

The obvious question, then, is: How do citizens make sure that private agendas and pursuit of profit do not erode consumer choice and even democratic expression?

Some people advocate government regulation as the best solution to this problem. Companies, they argue, should be banned by law from discriminating against content passing through privately operated networks. Networks should be required by law to be "neutral" in their handling of content—hence the term "net neutrality," coined originally by Tim Wu, a Columbia University law professor. In his influential 2010 book, *The Master Switch: The Rise and Fall of Information Empires*, Wu describes how technologies—from the telegraph and the telephone to radio and film—go through what he calls the "Cycle": new technologies start out open and revolutionary, then eventually get locked down by monopolists in ways that make them less appropriate or even inhospitable to uses that would be considered excessively political, edgy, provocative, or controversial. Yet Wu is also concerned about government interference in private networks in ways that violate citizen rights. He cites AT&T's role in the NSA's warrantless wiretapping as a clear example of how corporate power can be abused by an overreaching government.

Such abuse is most extreme, and indeed commonplace, in authoritarian countries where networks are by definition nonneutral. It has become standard practice for authoritarian regimes to require companies operating digital networks and platforms to censor and discriminate in an opaque and unaccountable manner. Furthermore, although networks and platforms carry out censorship at government direction, the lack of transparency about what is being censored and manipulated—and at whose behest—also makes it very easy for companies to discriminate in any number of ways that serve the interests of both the company and its executives. Citizens are thus vulnerable to abuse of their rights to speech and assembly not only from government but also from private actors. In democracies, it follows that citizens must guard against violation of their digital rights by governments and corporations—or both acting in concert—regardless of whether the company involved is censoring and discriminating on its own initiative or acting under pressure from authorities.

In late 2010 the Federal Communications Commission (FCC) issued rules stipulating basic net neutrality standards for the United States, with some caveats that free speech groups have criticized. Net neutrality proponents including Democratic Senator Al Franken believe it is a *pre-requisite* for Internet freedom, and ultimately as important as the First Amendment when it comes to ensuring a free and democratic discourse. Proponents tend to trust democratically elected government more than they trust private companies to act in the public interest.

Opponents of net neutrality argue that government interference of any kind with how private companies operate their networks runs counter to "Internet freedom." They also believe that as more and more Internet users turn to video and interactive gaming, network operators should be free to manage their traffic in whatever way is most efficient. They argue that market competition is the most effective way to keep companies honest. They also argue that excessive regulation stifles innovation, which ultimately hurts citizens, who are deprived of new products and services that cannot take root or thrive, thanks to regulations that inevitably fail to anticipate how technology will evolve.

There are certainly many strong reasons to be concerned about government meddling in private networks. But at the same time, market competition keeps companies honest only when there are ample choices and alternatives, and if the cost of switching services is not prohibitively high for most people. In the United States, that condition does not even come close to being met. Most Americans face very limited choices of broadband and wireless providers, especially in comparison to Europe, where regulations force competitive use of infrastructure.

As *The Economist* pointed out in December 2010 after the FCC issued rules on net neutrality that failed to satisfy purists on either side of the debate, the entire argument obscured "the failure in America to tackle the underlying lack of competition in the provision of Internet access." That, of course, would require much heavier political lifting in Congress: "Getting America's phone and cable companies to open up their networks to others would be a lot harder for politicians than prattling on about neutrality; but it would do far more to open up the net." As of 2011, there is no sign that Congress is serious about tackling this core problem of monopoly and pseudo-monopoly held by many wireless and broadband companies in many parts of the country. It is highly unlikely that any progress will be made in 2012, an election year in which all major telecommunications companies will contribute generously to congressional and presidential campaigns.

Responding to arguments by civic groups that net neutrality is the best way to support and nurture the growth of a robust digital commons, in 2010 Chile became the first nation to enshrine net neutrality into law, requiring that ISPs "ensure access to all types of content, services or applications available on the network and offer a service that does not distinguish content, applications or services, based on the source of it or their property."

In late 2010 the European Parliament and European Commission held a joint net neutrality summit, at which they agreed that it is permissible for companies to interfere with Internet traffic as long as the companies inform consumers that it is happening. But as net neutrality advocates point out, there remains a concern that as companies grow more comfortable with overtly interfering with customer communications in pursuit of their own interests, it will be easier for them to comply with government requests to conduct surveillance and censorship. Thus it was seen as an important step forward in April 2012 when the Netherlands became the first European country to make net neutrality the law when its parliament voted to ban mobile telephone operators from blocking or charging consumers extra money for using Internet-based communications services.

In many countries a lack of net neutrality makes censorship—whether by companies, government, or some mix of the two—much easier to implement and much less publicly visible, let alone accountable. In 2009 Jorge Bossio wrote a report on the Internet in Peru for the non-profit Association for Progressive Communications in which he pointed out that Telefonica del Peru had a clause in the customer agreement for its "Speedy home Internet service" that stipulated that "the customer is obliged to make use of content in accordance with the law, the present General Conditions, generally accepted moral values and good manners, and the public order," and that customers could be penalized for transmitting or posting content that was "false, ambiguous, inaccurate, exaggerated or untimely." In so doing, Bossio points out, the ISP had granted itself "powers that correspond to the Peruvian administrative and judicial authorities."

Lack of neutrality has other consequences around the world, which people in wealthy industrialized nations often fail to consider. Technologists in the developing world point out that if networks in the world's major Internet markets cease to be neutral, poor and developing nations will be at an even greater disadvantage in the global marketplace of online content and services. As the debates intensified in the United States and Europe in late 2010, Kenyan engineer and blogger Tom Makau wrote a post outlining what the outcome could mean for his continent: "With the lack of net neutrality, Africa-originated traffic will not be treated equally on the internet playing field as say traffic from Google or Microsoft," he wrote. "This is because the latter will be paying millions of dollars to have their traffic receive preferential treatment from the large data carriers. This, coupled with the fact that the continent produces slightly less than 2% of the total world traffic, will mean African content will be nearly invisible on the Internet."

MOBILE COMPLICATIONS

Even companies that support basic neutrality requirements for broadband Internet are taking a more relaxed view when it comes to mobile Internet access. Activists in the developing world who are waging an uphill battle for the right to use the Internet as a space for dissent and political mobilization were distressed by news in August 2010 that Google and Verizon had worked out a common position on net neutrality policy. The two companies agreed that "wireline broadband providers" should not be allowed "to discriminate against or prioritize lawful Internet content in a way that causes harm to users or competition." However, they also agreed that wireless broadband providers should be exempt from such regulation and ought to be free to manage and prioritize their traffic. Several months later, the FCC issued rules that generally reflected this compromise: though mobile carriers would be prohibited from blocking websites, they would be free to block applications or services unless those applications directly competed with providers' voice and video products, such as Skype. The rules allowed for some "network management," which includes prioritization of some services over others, and did not rule out "paid prioritization" of services, with the stipulation that carriers must be transparent about how they implement their traffic management. Net neutrality opponents and proponents were both highly critical with the compromise position.

Many net neutrality advocates are particularly concerned that although there is still hope that the broadband-based Internet in homes and offices can remain relatively neutral at least in democratic countries, policy and industry trends point to the likelihood that the *mobile* Internet *everywhere* will be much more constrained and manipulated, with limited room for the citizen commons. At the same time, there is a major demographic shift under way when it comes to global Internet use. The Internet is shifting from being a medium used primarily by people who live in economically developed, democratic countries to being a medium whose fastest user growth is in poor or developing countries—and in countries that are authoritarian, quasi-authoritarian, or very weak democracies. Much of the Internet's growth is happening through mobile phones, not through laptops or desktops connected to broadband.

It is therefore likely that Internet users around the world could be divided broadly into two classes: In one class are the wealthy, who have home broadband connections and who own PCs or laptops, and who are empowered by a much more open and neutral Internet. In the other class are the poor, the migrants, and the generally disenfranchised, who use the Internet mainly through highly controlled, easily tracked, non-neutral mobile devices. The situation is all the more insidious when we know that many of these companies are responding to government orders to censor and track users, in ways that are neither transparent nor accountable to the people who have come to depend on them.

What might a non-neutral wireless Internet service actually be like for consumers? In the developing world a picture is already emerging. In May 2010, Facebook launched a service called Facebook Zero in more than forty-five wireless markets, signing deals with mobile carriers to provide Facebook access free to subscribers. Now in a growing number of countries including Pakistan, India, Sudan, and Tunisia, mobile phone users must pay the standard data rate to browse the open Web but can access the special free Facebook service, at 0.facebook.com, for unlimited amounts of time. Furthermore, there is now a Facebook client available that works through a phone's SMS text messaging system, for phones that are not web-enabled, which allows even the cheapest phones without any access to the web to use Facebook.

For activist organizations seeking to reach and mobilize large numbers of people who do not have home broadband access or their own PC, this development has concrete implications. Imagine that a human rights organization in Sudan uses open-source blog-hosting software to set up its own website on a secure service, through which its members can completely control their identities and maintain anonymity, and makes sure that visitors to the site are not tracked. The organization also has a Facebook page, which risks shutdown by Facebook staff if the page's creators fail to comply with Facebook terms of service requiring the use of a real name. Facebook's privacy policies and features can also change without warning, exposing activists who take real risks of reprisal by authoritarian regimes when they use their real names on Facebook. In a country where open web browsing on the mobile phone costs significantly more than Facebook access, that organization is forced to work through Facebook—even if they prefer not to, due to security concerns—because the audience it most needs to reach has been driven by economic incentives away from the open web and onto Facebook.

The lack of neutrality in the mobile space also has troubling surveillance implications, because mobile phones are much more closely tied to the specific Internet user than PCs using broadband. Phone companies, which double as ISPs in many countries, can identify the user's device through its international mobile equipment identity (IMEI) number. The connection is identified by the SIM card number as well as the phone number. Even worse, many countries require registration of national ID or passport with the purchase of mobile phones (including

SIM card and IMEI numbers), guaranteeing close association of legally obtained phones with a real person's identity. Thus the mobile service provider can track who is accessing Facebook and when.

Mobile carriers are also experimenting with nonneutrality in the United States. In early 2011 the fifth-largest wireless carrier in the United States, MetroPCS, began offering a $40/month "no long-term commitment" service plan on one of its 4G phones, targeting lower-income customers. The service allowed unlimited talking, texting, web browsing, and YouTube access. Other premium multimedia services were available at additional cost, although some popular brand-name services like Skype and Netflix were excluded altogether. Apps from those services were blocked from being downloaded to the user's handset. Though the logic for the blockage is that these tools tend to use a lot of bandwidth, which costs the providers money, the precedent of blocking certain applications and not others is potentially a slippery slope. Because the MetroPCS service appears to violate the FCC's net neutrality rules, MetroPCS and Verizon joined hands in a lawsuit challenging those rules.

If net neutrality is defeated in the United States, what are the longer-term political implications? In addition to the reduction of choice, we also have a question of whether such policies will drive political discourse and activism to a handful of the biggest social media brands, whose services are likely to be chosen as preferred and free services targeting down-market customers. If I am staging a protest somewhere, and I use my camera phone to capture the scene of my friends being manhandled by police, if I want the video to be seen by the greatest number of people, these types of tiered and preferred services are going to force me to upload that video onto YouTube even if I want to use another video-sharing site whose terms of service I might prefer.

If the protest video I have posted to YouTube is removed by company staffers because it includes police violence and thus violates YouTube's community guidelines prohibiting graphic violence (there are numerous documented cases of activist videos being removed for this reason), or due to my accidental use of footage in which copyrighted music is playing in the background, which then triggers a takedown on copyright grounds (another frequent occurrence on social media sites), I am out of luck, particularly if I am trying to mobilize action and public attention around a time-sensitive matter and I do not have time to find a computer with a fixed Internet connection. I cannot use alternative video-sharing websites because my service allows me to use only YouTube.

BIG BROTHER APPLE

Apple's app censorship problem reaches well beyond China into unexpected places. In March 2010 Apple shut down, without notice, an iPad application for *Stern*, one of Germany's biggest magazines. It had published erotic content in the printed magazine, and that content along with all other magazine content was automatically duplicated in the iPad app. This content is perfectly legal in Germany, but because some pages of a specific issue were deemed in violation of Apple's app standards, the entire magazine was censored through the app store. Apple told another German magazine, *Bildt*, that it had to alter content if it wanted to keep its app. That year Apple also censored a cartoon version of James Joyce's *Ulysses*, which contained a few images of nudity, despite the fact that the app was already marked as containing adult content. The app was eventually reinstated, but only after a massive public outcry and widespread media coverage.

Apple also censors controversial political and religious content, including several apps featuring the work of Pulitzer Prize–winning cartoonist Mark Fiore because they "ridiculed public figures." His app mocking President Obama—containing the type of political humor frequently seen on television and in newspapers—was restored soon after Fiore won his Pulitzer. But apps by other cartoonists were not so fortunate.

Another satirist, Daniel Kurtzman, saw rejection by Apple of two apps meant to accompany his two satirical books, *How to Win a Fight with a Conservative* and *How to Win a Fight with a Liberal*. The apps were programmed to generate ridiculous insults, such as "May a commune of gay,

Marxist Muslim illegal immigrants open a drive-through abortion clinic in your church" and "Listen, you bongo-playing vegan, if ignorance is bliss, you must be one happy liberal." After repeated calls and e-mails, an Apple employee eventually explained that this app had been rejected because it involved insults directed at various groups of people. That the insults were completely ridiculous, and were actually meant to demonstrate that stereotype-based slurs are idiotic, was lost on Apple.

In response to widespread controversy over app censorship, in late 2010 Apple publicized its app review guidelines and established a review board so that developers would have a more systematic way to appeal decisions made against their apps. But censorship complaints continue, from both the Left and the Right of the political spectrum. In March 2011 conservative Christian groups erupted in outrage after Apple, responding to a protest petition signed by 150,000 people organized by Change.org, pulled an app created by the Christian group Exodus International that aimed to help "cure" gay men of their homosexuality. A similar app called the Manhattan Declaration, which condemned homosexuality as "immoral," had also been censored by Apple the previous November, also in response to a Change.org petition drive. The National Religious Broadcasters slammed Apple's decision as "a deliberate act of censorship of the religious speech and thought held by mainstream evangelical Christians in America."

Whatever one thinks about the content and message of the apps in question (I personally find them disturbing), the decision-making process behind their removal is troubling from a political and civil liberties standpoint. Though they can be swayed by public opinion if enough people succeed in galvanizing it, companies are also susceptible to the problem James Madison warned against in *Federalist* No. 10, when he warned that without adequate constitutional safeguards, unpopular speech or dissent could easily be silenced by an "unjust and interested majority." In Appledom, whistle-blowers and purveyors of highly controversial—yet constitutional—political views are thus every bit as vulnerable to takedown as the makers of the Exodus International and Manhattan Declaration apps.

Constitutional norms govern people's free speech rights in public spaces, but not in Appledom. Though these two apps are appalling to anybody who believes in gay rights, and though the public expression of homosexuality is offensive to some conservative Christians, the free speech rights of people with both sets of beliefs are protected by the First Amendment. Organizations promoting either side have a constitutionally protected right to publish websites. Even if no company agrees to host your web content, with sufficient technical know-how, anybody can set up a server in their own office or home and host the content themselves. Defending a home-served site from malicious denial of service attacks is increasingly difficult, but the point is that as long as the open web remains the primary platform for Internet content, it is technically possible for controversial groups to reach an audience—as long as they reside in a country where free speech is protected constitutionally and by the legal system. But if the bulk of the potential audience abandons the open web for closed, app-based systems, then political activists are increasingly hostage to the whims of corporate self-governance.

Fortunately for consumers, there are alternatives to Apple devices. The world's best-selling smart-phone platform as of mid-2011 was Google's open-source Android mobile operating system. Unlike Apple's iOS operating system, which runs only on Apple devices, Android can be adapted to run on any device made by any smart-phone or tablet PC manufacturer. Unlike Apple, which has a strict vetting system for iPhone and iPad apps, which are available only through its official app stores, Android apps can be created and distributed by anybody with the technical ability to do so, which means that Google does not exercise the same sort of political and cultural censorship that Apple imposes on its apps (although controls are still possible at the level of device-maker and local service provider). Google does carry out security vetting for Android apps for inclusion in its official Android App Market service, but unofficial apps are readily available through other third-party services.

As with many aspects of online and offline society, the flip side to more freedom can be reduced security—which is of course why security is government's most compelling justification for depriving people of freedom, and why finding the right balance between freedom and security is among the greatest challenges for any system of governance. This familiar problem now extends into the new dimension of mobile applications.

In mid-2010 the mobile security firm Lookout discovered an Android app providing customized "wallpaper" images that enable the user to personalize the appearance of their smart-phone display. This capability seemed innocuous enough, except that the app was also collecting the user's personal data and sending it to a "mysterious site in China." Google removed more than fifty apps from its Android Market in March 2011 and thirty additional apps in June. Security experts discovered that some of these apps provided seemingly useful or entertaining services but also contained malware that collected personal information from people's phones. Others were pirated imitations of legitimate programs issued by established companies, but reworked to include malicious code. Apple has, not surprisingly, sought to capitalize on this problem, calling Android apps "inferior." In the end, Google cannot control what a user installs on an Android device, because it does not control the device's hardware. Though Apple is not immune to inevitable security breaches, they are easier to contain because Apple controls both its hardware and its operating system, and the company can quickly push out patches and updates to all iPhones and iPads (as long as the user regularly syncs the device with the "mother ship" via iTunes).

Orwell's *1984*—which Apple once used so cleverly to market its products as rebellious and empowering—is the story of a government that manipulates citizens' fear of war and chaos to justify depriving people of freedom of speech and choice. As it has aged, Apple has become more like Big Brother than anybody could have imagined three decades ago.

Of course, the difference between totalitarianism and democracy hardly amounts to a choice between security and criminality. The distinction between democratic governance and anarchy is that government is empowered by citizens to defend them from criminals and malicious attacks, domestic and foreign. The point of "consent of the governed" is that citizens trade absolute freedom to do anything and everything they please regardless of its impact on others, for a certain degree of security for themselves, their families, and their communities. This trade-off works reliably in the citizens' interest, however, only if government is held accountable—not only by a competitive political system and a strong, independent legal system, but also by constitutional constraints that guarantee respect and protection of citizens' most basic rights, which include free expression and assembly.

In governing our access to and use of applications, Apple provides a valuable service by shielding us from malicious criminals. But it also shows troubling disregard for our political rights as citizens. Like all sovereign powers, Apple and other companies that are failing to protect and respect our rights to free expression and assembly are unlikely to change their attitudes or actions unless and until their customers and users force change upon them—just as citizens have forced change on governments that ultimately realized that their survival required it.

CHAPTER 9

Do No Evil

In April 2011, Mike Lazaridis, co-CEO of Research in Motion (RIM), maker of BlackBerry, sat down for an interview with the BBC's Rory Cellan-Jones. The main purpose of the interview was to discuss RIM's new tablet device. But at the end, Cellan-Jones asked Lazaridis to give an update on RIM's "arguments with the Indian government and various other governments in the Middle East." He wanted to know whether they were "anywhere near being sorted out."

Lazaridis replied, "That's just not fair, Rory." A woman (presumably a public relations person) interjected off-camera, "We're really up on time. Is there any other question you want to ask?"

Cellan-Jones's question was, by elementary standards of journalism, a reasonable and obvious one. In August 2010, the United Arab Emirates and Saudi Arabia had threatened to ban BlackBerry services unless RIM agreed to allow a satisfactory level of government access to all communications sent through RIM devices within the country. Governments can easily command local mobile service operators to share communications sent through BlackBerry's consumer devices sold to individuals. But RIM's BlackBerry Enterprise service, used by companies, governments, and other organizations, is set up in such a way that even RIM cannot access the contents of e-mails sent through the customer's own designated servers. India soon followed suit with its own demands for access. Though negotiations with the Indian government have dragged on inconclusively, RIM reached agreements with the governments of Saudi Arabia and the UAE so that service would not be disrupted. The company has not disclosed any details of those agreements.

> Thus Cellan-Jones persisted: "Why is that not a fair question to ask?"
>
> Lazaridis replied: "Because you said, you implied we have a security problem. We don't have a security problem."
>
> Cellan-Jones: "Well, you have an issue."
>
> Lazaridis: "No, we don't. We've just been singled out, right, because we're so successful around the world. It's an iconic product, used by business, it's used by leaders, it's used by celebrities, it's used by consumers, it's used by teenagers. I mean, we were just singled out. Just because of our success."
>
> Cellan-Jones: "But has that been sorted out now?"
>
> Lazaridis: "Look, we're dealing with a lot of issues and we're doing our best to deal with the expectations that we're under."
>
> Cellan-Jones: "And you're confident that—we've got a lot of listeners and viewers in the Middle East and India. You can confidently tell them that they're going to have no problems with being able to use their BlackBerries and you being able to give them assurance that everything is safe and secure?"
>
> Lazaridis: "It's over. The interview is over. Please. You can't use that, Rory, that's just not fair. Its not fair."
>
> Cellan-Jones: "Well . . ."
>
> Lazaridis: "C'mon, we've dealt with this. This is a national security issue. Turn that off."

Lazaridis apparently considers himself a victim of unfair media coverage. But his company holds far too much power over people who depend on its services to justify such a victim mentality—or to dismiss summarily such concerns on the part of journalists, its own customers, or the general public. RIM is cooperating with governments and possibly compromising user security, yet its executives emphatically refuse to discuss these issues. Other companies—particularly Yahoo with its experiences in China—have learned the hard way that this sort of "stick your head in the sand" response to admittedly tough ethical dilemmas and regulatory problems is unlikely to serve RIM well in the long term.

CHINESE LESSONS

When Chinese state security agents detained him on November 23, 2004, Shi Tao was a bespectacled and clean-cut thirty-six-year-old journalist, married with a young son. When he finishes his ten-year jail sentence in 2014, he will be a broken man. His wife divorced him soon after his ordeal started. Hard labor at a prison factory in the early years of his sentence gave him respiratory problems. He has an ulcer and a heart condition. In 2007, the last time his mother, Gao Qinsheng, was allowed out of China, she told journalists that her son's mental state had "deteriorated."

Shi's ordeal stemmed from an editorial meeting on April 20, 2004, at *Contemporary Business News*, a newspaper in the city of Changsha, Hunan province. At the meeting, a senior editor gave a verbal summary of an official government document issued to media organizations nationwide. Authorities were nervous about possible protests in the run-up to the June 4 anniversary of the 1989 Tiananmen Square crackdown, and the document contained detailed instructions to news organizations about what should and shouldn't be reported over the coming months. Shi took notes. That night, working late in the office, he logged into his Yahoo China account (huoyan-1989@yahoo.com.cn) and summarized what his editor had said at the meeting. He e-mailed his summary to a New York–based editor of a prodemocracy website, Democracy Forum, requesting that it be published immediately under the name "198964."

Two days later, on April 22, the Beijing state security bureau issued a "Notice of Evidence Collection" to Yahoo China, which was at the time a subsidiary of Yahoo Holdings (Hong Kong), Ltd. The order requested "email account registration information for huoyan-1989@yahoo.com.cn, all login times, corresponding IP addresses, and relevant email content from February 22, 2004 to the present." Beijing-based employees of Yahoo complied with the request on the same day.

After being detained without charge or access to a lawyer for nearly a month, in mid-December Shi was formally arrested and charged with "leaking state secrets." His trial on March 11, 2005, lasted two hours. The guilty verdict and ten-year prison sentence came a month and a half later.

The outside world did not learn about Yahoo's role in Shi Tao's conviction until September 2005 when Reporters Without Borders published a copy of the court verdict, obtained and translated by the Dui Hua Foundation, a China-focused human rights group. Among the evidence against Shi, court documents listed "account holder information furnished by Yahoo Holdings (Hong Kong) Ltd." Over the course of the next several months, human rights groups would uncover three more cases in which Yahoo had supplied evidence that contributed to the conviction of political dissidents.

The company's initial response to critics focused on the fact that Yahoo employees in China—most of them Chinese citizens—had no choice but to comply with Chinese law. Cofounder Jerry Yang said at the time that although he felt "horrible" about what had happened, "we have no way of preventing that beforehand. . . . If you want to do business there you have to comply."

Yang's attitude made legal sense. Yahoo's defenders—including prominent Silicon Valley–based entrepreneurs, bloggers, and business journalists—pointed out that Yahoo's Chinese employees risked arrest themselves if they had refused to comply with the Chinese state security bureau order. Furthermore, Yahoo's China-based employees acted exactly as the Yahoo China e-mail user terms of service promised they would. When Shi Tao signed up for an e-mail account on yahoo.com.cn, he could have done so only if he clicked "agree" on the terms of service, which ask the user not to commit a list of actions, including "damaging public security, revealing state secrets, subverting state power, damaging national unity," etc. The same document to which Shi had technically consented (regardless of whether he actually read or understood it) acknowledged that he understood that his information would be disclosed to authorities if required to do so by law.

Legally, Yahoo was off the hook. Ethically, it was not. A man's life was ruined, and Yahoo had played an undeniable part in ruining it. Human rights groups and others argued that companies have broader moral obligations beyond local and home-country law. After all, the parts of Chinese law that define "crime" to include peaceful political speech themselves violate international laws and covenants, such as the Universal Declaration of Human Rights. As Congressman Tom Lantos put it at a dramatic congressional hearing lambasting executives from Yahoo as well as Google, Microsoft, and Cisco in February 2006, "If the secret police a half century ago asked where Anne Frank was hiding, would the correct answer be to hand over the information in order to comply with local laws? These are not victimless crimes. We must stand with the oppressed, not the oppressors."

In the spring of 2007, Shi Tao's mother, Gao Qinsheng, and Yu Lin, the wife of Wang Xiaoning, another Chinese dissident who had been convicted with Yahoo's help, sued Yahoo in the US District Court of Northern California. They accused the company of having "unlawfully accessed and used, and voluntarily disclosed, the contents of the intercepted communications to enhance their business in China." Yahoo filed a motion to dismiss, arguing not only that the case fell outside of the court's jurisdiction, but also that the company could not be held liable: it was bound by Chinese law to comply with a lawful request by Chinese authorities—and acted strictly according to terms of use, to which the plaintiffs had technically consented by clicking on that "agree" button.

The World Organization for Human Rights USA, which represented the family members in their lawsuit, retaliated by seeking court discovery of internal company documents related to the case. Congressman Lantos summoned Yahoo cofounder Jerry Yang for a second congressional hearing in October 2007, calling Yang and his colleagues "moral pygmies." After sitting through emotional, tearful testimony by Shi Tao's mother, Gao Qinsheng, Yang bowed solemnly to her three times and said, "I want to personally apologize." Soon thereafter, Yahoo settled with the families of Shi Tao and Wang Xiaoning for an undisclosed sum.

It had taken two years of being pummeled by Congress, human rights groups, the media, and shareholders before Yahoo finally shed its head-in-the-sand, lawyer-driven posture and actually took moral responsibility for what had happened. The company set up an internal Business and Human Rights Program, a fellowship at Georgetown University to support research on the Internet and human rights, and a Yahoo Human Rights Fund administered by Chinese human rights activist Harry Wu to "provide humanitarian and legal assistance to persons in the People's Republic of China who have been imprisoned or persecuted for expressing their views using the Internet."

Yahoo was certainly not the only company that found itself in hot water with Congress and human rights groups for its actions in China, but it is important to note that other companies, particularly Microsoft and Google, did manage to learn from Yahoo's mistakes and avoided repeating the worst of them. They made different business decisions in China that were not without controversy, but that at least helped them avoid complicity in the arrest of Chinese dissidents. Their experiences prove that the classic argument that many executives make—that they have no choice but to comply with local government demands if they want to be in the market at all—is both overly simplistic and irresponsible. There are many different ways to "be" in a market. The challenge is to consider the specifics of "how" before committing the company to a no-win situation for all concerned, as Yahoo did by rolling out an e-mail service inside China.

Though Microsoft chose not to launch a Chinese version of its e-mail service, it did decide to enter China with a censored search engine and a censored blogging platform. The year Microsoft launched MSN Spaces in China, 2005, was also the year the Chinese blogosphere exploded, going from around half a million Chinese bloggers in January to more than five million in less than a year. MSN Spaces has been credited with fueling that growth. On the other hand, 2005 was also when the government began to set up its "self-discipline" system for controlling and censoring social media through the companies themselves. To keep from getting blocked by the "great firewall," Microsoft had to agree to police its Chinese blogs for politically sensitive content.

Microsoft's collaboration with Chinese corporate censorship requirements hit the international spotlight in early January 2006 after MSN Spaces suddenly deleted—without any warning—a blog written by the Chinese journalist Zhao Jing, writing under the pen name Michael Anti. Zhao was one of China's edgiest journalistic bloggers. He had angered authorities by writing about a government crackdown against the *Beijing News*, a new tabloid with a national reputation for exposing corruption and official abuse. The takedown sparked an immediate outpouring of rage in the American technology blogosphere, particularly after Microsoft admitted to the *New York Times* that MSN Spaces staff had deleted Zhao's blog "after Chinese authorities made a request through a Shanghai-based affiliate of the company."

The public outcry was so strong in the United States that by late January, Microsoft had changed its procedures for handling Chinese government censorship demands. Among those called by Congressman Lantos to testify alongside Yahoo executives in February 2006 was Jack Krumholtz, Microsoft's associate general counsel. He outlined the company's attempt to provide transparency while still complying with Chinese censorship requirements: from now on, blog posts would be blocked only in response to a "legally binding notice from the government" (rather than a casual phone call, e-mail, or other non–legally binding communication) and access to that content would be blocked only to users in the country issuing the order (a technique known as "geo-filtering"). Users would be notified why their content had been blocked.

Just as Yahoo and Microsoft came under attack in the US media for their actions in China, Google chose January 2006 to launch its Chinese-language search engine, Google.cn. Alongside Microsoft and Yahoo, Google was also excoriated by Congressman Lantos for subjecting its search engine to Chinese government censorship. But Google executives pointed out that the company had drawn its own ethical line more stringently than either Yahoo or Microsoft: not only had Google decided to forgo a local version of Gmail to avoid inevitable involvement in cases like Shi Tao's, but also, even though Microsoft had compromised its blog-hosting platform to be in China, Google had decided not to roll out Chinese versions of its own blogging platform, Blogger, or other content-sharing platforms, such as YouTube. Though Google was willing to censor search results for a time (until eventually pulling out of the Chinese search market in 2010), the company was never willing to censor or delete the work of individual Internet users.

The companies' decisions to comply with—and thus help to legitimize—parts of the Chinese Internet censorship system were highly controversial. Google executives have been honest about the fact that the company was divided internally about the best way to proceed. China's liberal bloggers, trying doggedly to carve out a space for independent discourse inside China, on the whole tended to support the decision by Microsoft and Google to provide service to Chinese users, which, while compromised in terms of censorship, maintained red lines on surveillance that these companies vowed not to cross. Even Zhao Jing, whose own blog had been deleted by MSN Spaces, wrote in February 2006 that the Chinese blogosphere would be better off if Microsoft and Google stayed than if they refused to make any compromise and pulled out. "Companies such as Microsoft and Google have provided Chinese netizens with much freedom of information over these years," he wrote. "China also needs these American companies." Yahoo, however, was another story: "A company such as Yahoo which gives up information is unforgivable."

Making the right decisions is difficult in situations that are often not black-and-white, as the China situation clearly shows. Companies do not have sufficient in-house human rights expertise or experience to anticipate and prevent all possible risks their users face and the problems they may encounter everywhere they do business. For this reason, in late 2008 Yahoo, Google, and Microsoft helped launch the Global

Network Initiative (GNI), a multi-stakeholder initiative through which companies work with human rights groups, academic experts, and socially responsible investors to promote implementation of baseline standards of free expression and privacy. (Chapter 11 will describe the GNI in greater detail.)

After signing on to the GNI principles and making a public commitment to uphold basic human rights, Yahoo has so far avoided direct complicity in further dissident arrests in China and elsewhere in the world. But that does not mean that GNI member companies have escaped all such future problems. Nor can they be expected to, given the complexities and dilemmas they will continue to face.

FLICKR FAIL

For any company providing commercial services whose purpose was meant to be something other than political activism, business concerns can easily come into conflict with human rights concerns—usually at unexpected moments, with unexpected consequences. That is precisely what happened with Yahoo's photo-sharing service, Flickr, and Egyptian activists in early 2011. For several years preceding Egypt's revolution, a number of activists worked to expose state security agents, outing them on the Internet. One of these activists, Hossam el-Hamalawy, built up a large collection of photos of Egyptian state security agents taken by protesters attending demonstrations and posted them on Flickr. El-Hamalawy even wrote an Arabic-language guide to Flickr for activists, praising the service's usefulness for political activism. Then in March 2011, when activists stormed state security offices around the country soon after the Mubarak regime fell, el-Hamalawy found a disk in state security's Cairo headquarters containing hundreds of photographs of agents. As part of his long-standing agent-outing project, he uploaded many of those photos to Flickr. The idea was to ensure that these people would not become part of Egypt's new government and that those known to have engaged in torture would be identified and brought to account. To his shock and outrage, Flickr staffers deleted them. Why? Because he had violated one of Flickr's community rules: "Don't upload anything that isn't yours."

This was not a case of Egyptian police telling Flickr to take down material. It was not a matter of employees being forced to do something, or having no choice under duress. No law was broken, nobody at Flickr risked going to jail, and the company didn't risk punitive action by any government if it kept el-Hamalawy's photos online. This was a case of Flickr enforcing its own internal rules—which all users technically agree to when they sign up for the service.

Ebele Okobi-Harris, director of Yahoo's Business and Human Rights Program, explained in a blog post and on a panel the next month at the South by Southwest Interactive conference that Flickr's staff had removed those photos because a user reported them for violating community rules. "Flickr was set up as a community for people who love photography to share their photographs," she said. "In this particular case, we had someone who wanted to use Flickr, not for photographs that he took, but for photographs that he found somewhere else. The community rules are about sharing your own content. You can't upload photos that are not your own." But as many people immediately pointed out to her, one can find plenty of examples throughout the site where people have uploaded photos that clearly are not theirs. Somebody called "Lee" replied in a comment on her blog, "Thirty-five thousand photos tagged as 'not mine' on the site and these are the ones you go after. That's pathetic."

Veteran activist Gilles Frydman, who had brought the issue to public attention a few days previously, challenged Okobi-Harris outright at the conference: "So should I tell activists to just leave Flickr?" She admitted that Flickr does not claim to be "a place for everyone"—that it is set up for a certain purpose and is managed in a certain way, and that some activists may reasonably conclude that its rules and governance methods are not compatible with their aims. Also in the room was photographer Thomas Hawk, who reacted angrily to the exchange. "The fact remains that what Yahoo and Flickr did does NOT support free expression as a fundamental human right," he wrote. "It does NOT support the idea that technology and more specifically Flickr should be used to create social change."

Okobi-Harris and other Yahoo executives were honest enough to admit that the service is not managed perfectly; that el-Hamalawy's case might have been handled with greater sensitivity even if the final outcome were the same; and that the company needs to think more about how to avoid inadvertently harming activists while still empowering staff to enforce terms of service and community guidelines. But it is also a fact that neither Flickr nor any other commercially operated service is required to uphold the First Amendment for American users or Article 19 of the Universal Declaration of Human Rights, which guarantees the right to free expression, for its global users. And as el-Hamalawy's case clearly shows, the company is comfortable restricting speech and images that are considered "protected speech" according to globally accepted human rights norms. Similarly, the owner of a private building has the right to prevent visitors from using the space to hold a political rally, despite the fact that the right to organize peacefully in public spaces is protected by international human rights norms. Flickr is not a public space like the National Mall in Washington, DC, or Tahrir Square in Cairo; it is more like a shopping mall or a food court, where the private owner of the space retains the right to decide who can or cannot occupy its space and under what terms.

BUZZ BUST

Companies sometimes construct these quasi-public spaces in ways that seem like fun houses to their own employees but end up as houses of horror for other people who live and work in very different contexts. Take, for example, what Google inadvertently did in early 2010 to a woman living somewhere in the Midwest, who calls herself "Harriet Jacobs" (not her real name). As a child she fled an abusive home. In her mid-twenties Harriet fled a husband who raped her regularly. As part of an effort to come to terms with her own life and help others, she began writing a blog called "Fugitivus," an open diary and virtual support group for herself and anybody else out there on the Internet who found it useful.

Harriet managed all of her e-mail accounts (some tied to her real identity and others tied to her online persona) through Gmail, the free service run by Google. Gmail has a particularly handy feature that enabled her to automatically send all of her ex-husband's threatening e-mails, along with mail from other hostile people in her online and offline lives, into a folder labeled "People I Don't Like." She also used a service called Google Reader, which allows users to manage subscriptions to all their favorite blogs and news websites, and enables users to share and discuss them with other people they specify—*if* they choose to. The service also has settings that allow users to decide whether they want to share their "likes" and recommendations with the general public, with a designated circle of contacts, or with themselves only. This Gmail/Google Reader combination worked well for Harriet, managing her e-mail as well as her daily intake of news and blogs.

Then on February 11, 2010, she logged in to find a nightmare unfolding online, with real consequences for her physical safety. Without warning to people who use Gmail and Google Reader, and without giving them the ability to "opt out" from the get-go, Google launched a new social networking service called Buzz. The idea behind Buzz was to create an instant social network from the most frequent contacts in a person's Gmail account, plus their buddies in Google Reader as well as Google's instant-messaging service, GTalk. As a result, by default, Buzz exposed all of her frequent Gmail contacts and the people in her Google Reader network to one another.

In creating Buzz, Google's programmers and product managers decided that it made sense to combine people's frequent Gmail contacts and Google Reader buddies as the starter members for a new online social network, which Gmail and Google Reader users automatically became members of. It may have seemed smart and harmless to them, but as Harriet put it in an angry blog post, "My privacy concerns are not trite. They are linked to my actual physical safety."

Unlike websites and online communities formed around electing certain candidates or passing certain laws, Harriet's "Fugitivus" is not directly engaged in politics. Yet in a broader sense, "Fugitivus" is very political. A lot of what Harriet writes on her blog deals with working assumptions made by various public and private institutions in America about the nature of rape, abuse, and male-female relations inside and outside of marriage, which in turn can contribute to the victimization or empowerment of women, depending on how problems are handled. If that isn't political, what is? Platforms that help women to redefine themselves and their lives, and that might make them more confident in confronting the status quo, are profoundly political. Silicon Valley CEOs talk regularly about how their products are empowering activists to speak truth to power all over the world. But if people like Harriet who live in democracies do not feel safe using the Internet for difficult and uncomfortable conversations, women and girls everywhere will be worse off in many ways, including politically.

A few days after Harriet was exposed on Buzz, Google issued a public apology after being swamped with user complaints and negative news reports. Google's policy team explained that the company was scrambling to rectify what they realized was an egregious breach of user privacy and trust. The company's programmers quickly rolled out changes so that Buzz would no longer be automatically connected to people's Gmail. They also made it easier for people to control their settings or disable Buzz completely. In the end, executives made a convincing case that the breach—with potential life-and-death consequences for vulnerable users around the world—was not deliberate. It was the result of shocking failures in management and product vetting that one may expect from a small start-up but that are inexcusable when they happen in a company as large and powerful as Google, with responsibility for the privacy of hundreds of millions of users around the world.

Somehow the company had allowed Buzz to be launched into the accounts of all Gmail and Google Reader users after being tested only by Google employees. The Buzz development team had completely by-passed Google's normal trial and testing procedures that would have in-volved a more diverse range of users: people who are more likely to have brought up the issues that were so glaringly obvious immediately after the launch. The potential danger for people like Harriet somehow had not occurred to the engineers working on their sunny campus in Moun-tain View, California—apparently because Harriet's concerns are so alien to their own life experience. They had not sought out anybody remotely like Harriet to help them test the new service. Google's fa-mous motto may be "don't be evil." Its internal product development processes heard no evil and knew no evil, but inadvertently did evil—even though the company had signed on to a pledge to uphold free speech and privacy.

In response to complaints filed by privacy groups, in April 2011 the Federal Trade Commission issued a ruling that required Google to im-plement a comprehensive privacy program and submit to independent privacy audits for the next twenty years. The company is also required to obtain users' explicit consent before including them in any new service or to share their information in new or unexpected ways. As Google scrambled to deal with the fallout from Buzz in early 2010, Facebook was also scrambling to undo at least some of the damage after it altered privacy defaults without sufficient concern for its most vulnerable users.

PRIVACY AND FACEBOOK

As protests mounted in reaction to Iran's presidential elections on June 12, 2009, Facebook actively encouraged members of the pro-opposition Green Movement to use the social networking platform. By mid-June, more than four hundred members of Facebook's fast-growing Farsi-speaking community volunteered their time to create a Farsi version of Facebook. Thanks to efforts by Facebook enthusiasts all around the world, the platform has been made accessible in seventy languages—including many languages spoken in countries where regimes are known not to tolerate dissent.

Then in December 2009, Facebook made a sudden and unexpected alteration of its privacy settings. On December 9, to be precise, people who logged in got an automatic pop-up message announcing major changes. Until that day, it was possible to keep one's list of Facebook "friends" hidden not only from the general Internet-surfing public but also from one another. That changed overnight without warning. An array of information that Facebook previously had treated as private, suddenly and without warning became publicly available information by default. This included a user's profile picture, name, gender, current city, what professional and regional "networks" one belonged to within Facebook, the "causes" one had signed on to support, and one's entire list of Facebook friends.

The changes were driven by Facebook's need to monetize the service but were also consistent with founder Mark Zuckerberg's strong personal conviction that people everywhere should be open about their lives and actions. In Iran, where authorities were known to be using information and contacts obtained from people's Facebook accounts while interrogating Green Movement activists detained from the summer of 2009 onward, the implications of the new privacy settings were truly frightening. Soon after the changes were made, an anonymous commenter on the technology news site ZDNet confirmed that Iranian users were deleting their accounts in horror:

> A number of my friends in Iran are active student protesters of the government. They use Facebook extensively to organize protests and meetings, but they had no choice but to delete their facebook accounts today. They are terrified that their once private lists of friends are now available to "everyone" that wants to know. When that "everyone" happens to include the Iranian Revolutionary Guard and members of the Basij militia, willing to kidnap, arrest, or murder to stifle dissent, the consequences seem just a bit more serious than those faced from silly pictures and status updates. I realize this may not be an issue for the vast majority of American facebook users, but it's just plain irresponsible to do

this without first asking consent. It's even more egregious because Facebook threw out the original preference (the one that requested facebook keep the list of friends private) and replaced it with a mandate, publicizing what was once private information—with no explicit consent. If given the choice to remain a facebook user with those settings, or quit, my friends would have quit rather than risk that information being seen by the wrong people. Instead, Facebook published it anyways. It's a betrayal of trust for the sake of better targeted advertising.

The global outcry over the exposure of people's friend lists in December 2009 was so strong that within roughly a day after the dramatic change, Facebook made an adjustment so that users could once again hide their friend lists from public view. People's friends could still see one another, however, and there was no way to hide them. Everybody's "causes" and "pages" were still publicly exposed by default, another serious vulnerability for activists. People kept complaining—many by creating protest groups within Facebook itself, where hundreds of thousands of people from all around the world posted angry messages. The groups had names like "Facebook! Fix the Privacy Settings!" and "Hide Friend List and Fan Pages! We Need Better Privacy Controls!" and "We Want Our Old Privacy Settings Back!" Scrolling through these pages, you see people posting from all over the world, with large numbers of Arab, Persian, Turkish, Eastern European, and Chinese names.

Eventually Facebook fixed this problem as well, adjusting the privacy options so that information about what pages users follow, or groups they have joined, can be made private. Meanwhile, however, lives of people around the world had been endangered unnecessarily—not because any government pressured Facebook to make changes, but because Facebook had its own reasons and did not fully consider the implications for the service's most vulnerable users, in democratic and authoritarian countries alike. As privacy advocate Kaliya Hamlin put it, "What if you were HIV positive and followed a page for a group that provided services to people like you? What if you had not told your work colleagues about this health condition? Now with your pages being public, your health status is completely public."

People on the American political fringes also felt violated. One example was that of people advocating to legalize polygamy, which is illegal in the United States and is likely to remain so: according to a 2009 Gallup poll, 91 percent of Americans "consider polygamy wrong." Like many people who go on national TV to exercise their First Amendment right to express highly unpopular ideas, polygamy advocate Mark Henkel gets a lot of death threats, not from authorities but from random people who find his views reprehensible. For this reason, Henkel was a latecomer to online social networking. He was worried about security and the need to protect the privacy of his supporters.

But many members of his pro-polygamy organization urged him to take the plunge. He was reassured by those who pointed out that if you sign up for a personal account, everybody who has joined your friend list can be kept private—from the general web-surfing public as well as from one another. Other members of his organization made the case that if Facebook was safe enough for opposition activists in Iran, it should work for the American polygamist movement. "Facebook was selling themselves as a vehicle for democracy," Henkel told me in a phone interview. "We trusted that if the Iranians were safe, we would be, too." So he took the plunge in August 2009.

Then in December, Facebook suddenly changed its privacy settings. Needless to say, the pro-polygamy community—along with Facebook's Iranian users—was horrified. "It was a breach of trust and an immoral act," Henkel said. "If your privacy isn't protected, you can't have civil disobedience. We trusted that Facebook was keeping our information private but then they suddenly chose to do otherwise. I don't see how they can ever be trusted again."

The problem is that political activism around the world increasingly depends on what are, in effect, the cyberspace equivalents of food courts and shopping malls. Citizens face the risk that these spaces will be turned into opaque extensions of state power in many countries. Yet even when companies do avoid government interference, they are at best benevolent dictatorships: creating and enforcing their own rules as they see fit. Because they obtain our de facto consent when we create an account and click "agree" to the terms of service, our protests against their rules or how they choose to enforce them are considered invalid—and no law in any major democracy will back us up if we insist otherwise.

CHAPTER 10

Facebookistan and Googledom

In May 2010, Hong Kong–based university professor and communications scholar Lokman Tsui decided to delete his Facebook account. In a blog post explaining his decision, he likened Facebook to a country run by an authoritarian, paternalistic government that claims to be acting in its people's best interest:

> Allow me to make a wild analogy, one I believe is not entirely out of left field. Many people know that there is censorship in China. Many people also tell me that 1) the poor Chinese must feel really repressed or 2) they must be okay with it. But if that's the case, who in their right mind can be okay with censorship? They must be brainwashed.
>
> Ask yourself this: if I decide not to leave Facebook, yet I know they do not care at all about my privacy, what does that mean? How is that different from the people who continue to use the Internet in China day in day out despite the prevalent and prolific practices of censorship? This is not a rhetorical question. Of course I realize Facebook is not the Chinese government, but I do think there are similarities between them, in kind although perhaps not in degree.

Leaving Facebook is so many magnitudes easier—physically, economically, emotionally—than it is for the average Chinese citizen to leave China and start a whole new life in another country. A physical government's power over the individual is not in any way comparable to the power that any Internet company holds over any person. Still, Tsui made an important point. Hundreds of millions of people "inhabit" Facebook's digital kingdom. Call it Facebookistan.

By mid-2011 Facebook had 700 million users. If it really were a country, it would be the world's third largest, after India and China. The social network may have started out in Mark Zuckerberg's Harvard dorm room as a platform for college students to flirt with one another, but it is now a world unto itself: an alternative virtual reality that for many users is now inextricably intertwined with their physical reality—and one that is often celebrated as a platform not only for personal expression but for political liberation.

Facebook's motto is "Making the world open and connected." In his best-selling book *The Facebook Effect*, charting the company's origins and growth through the end of 2009, journalist David Kirkpatrick describes Zuckerberg's deep and long-standing belief in what he calls "radical transparency": the idea that humanity would be better off if everybody were more transparent about who they are and what they do. Anonymous online speech runs directly counter to this vision. Zuckerberg tells Kirkpatrick, "The days of you having a different image for your work friends or co-workers and for the other people you know are probably coming to an end pretty quickly. . . . Having two identities for yourself is an example of a lack of integrity."

This vision of the world—and Facebook's role in shaping it—is deeply embedded in how Facebook's top executives, developers, marketers, and programmers think about the service and its purpose. It is their ideology. It is the foundation upon which the laws of Facebookistan are constructed. The terms of service, to which every user much click "agree" to create an account, require that all inhabitants of Facebook use their real names. The sovereign rulers of Facebookistan enforce this "real-ID" policy. When discovered, accounts using pseudonyms or fake identities are punished with account suspension or deactivation. This internal governance system spans physical nations, across democracies and dictatorships. It influences people's ability to communicate not only through Facebook itself, but also through a rapidly expanding universe of other websites and services that are increasingly integrated with Facebook.

DOUBLE EDGE

By mid-2010, an estimated 3.4 million Egyptians were on Facebook, making Egypt the top Facebook-using country in the Arab world. In the spring of 2008, the Egyptian government—which had been led for nearly three decades by the same president, Hosni Mubarak—first felt the force of Facebook activism when young people in Cairo used the social networking platform to organize protests involving more than 60,000 people against rising food prices.

Then in June 2010 a young man named Khaled Said was brutally murdered by police in Alexandria, in retaliation, his family believed, for incriminating video of them that he planned to post on the Internet. After photos of his mutilated body in the morgue made it onto the Internet, several activists including a young Google executive named Wael Ghonim set up a Facebook page called "We Are All Khaled Said," using fake names to protect themselves from potentially meeting the same fate as their hero. The group organized a series of protests called "silent stand against torture" involving first hundreds then thousands of people in cities across Egypt. More than 1,000 people showed up for Said's funeral. More than 8,000 people attended one protest in Alexandria.

The day before one long-planned Friday of protest—which happened to be Thanksgiving Day back at Facebook's Palo Alto headquarters— the Khaled Said page hit its peak of activity as more people joined, members traded information, and organizers sent out updated instructions. Then suddenly, without warning, the page disappeared from view. Its creators received a notice from Facebook staff that they had violated terms of service that require administrators of pages to use their real identities—and furthermore, that accounts of people not using their real names, when discovered, would be shut down.

The page's creators were fortunate to know people working in Silicon Valley and for international human rights groups, who contacted Facebook executives. The Khaled Said page was restored in less than twenty-four hours, but only after administrative rights for the page were handed over to another person willing to verify her true identity with Facebook staff. After the revolution, that person felt safe enough to reveal her name publicly as Nadine Wahab, an Egyptian woman living in Washington, DC. After taking over responsibility for the Khaled Said page in November 2010 when the group's members were particularly vulnerable to arrest—and the same kind of police brutality that had killed Khaled Said in the first place—Wahab said she found Facebook's absolutist attitudes toward anonymity exasperating. "These guys are techies," she told me in December. "I don't think they understand the implications that their rules and procedures have for activists in places like Egypt."

If anybody needed a reminder of how dangerous the situation was, just one week after the Khaled Said deactivation incident, a thirty-year-old man named Ahmed Hassan Bassyouni was brought before a military tribunal. They sentenced him to six months in prison. Why? Because he created a Facebook page dedicated to advising people on the application process for joining the Egyptian military. It seemed harmless enough that a local radio station interviewed him about it— something they would not dare to do if he was running a page, for example, about military corruption. As his lawyer pointed out, the information on Bassyouni's page was all publicly available from official sources and regularly published in newspapers. Still, he was accused of "spreading military secrets over the Internet without permission." Unlike Wael Ghonim, who was careful to hide his identity while running the Khaled Said protest group, it had not occurred to Bassyouni that he would get in trouble for what he was doing. He made the mistake of being open about his real identity on Facebook, and paid for it dearly.

"Once these kinds of things happen in a community where brutality is constant," Wahab told me in December 2010, "Facebook no longer feels like a safe place." The problem is, "there are no other alternatives now. If you want to organize a movement the only place to do it effectively is on Facebook, because you have to go where all the people are. There needs to be a mechanism that enables us to do this kind of work. Either Facebook is going to get it, or we're going to be playing cat and mouse." Fortunately for Egypt's activists, the story ended well, at least in the short term, with the fall of the Mubarak regime. Wael Ghonim was soon lavishing praise on Mark Zuckerberg for having created the world's greatest organizing tool for freedom and democracy.

INSIDE THE LEVIATHAN

Members of Facebook's management team are adamant that the real-name requirement is key to protecting users from abusive and criminal behavior. Tim Sparapani, who worked for the American Civil Liberties Union before becoming Facebook's public policy director, explained it to me this way: "Authenticity allows Facebook to be more permissive in terms of what we can allow people to say and do on the site." The worst behavior on Facebook, he said, is committed by people known as "trolls" who try to hide behind fake identities to get away with abusive behavior that they would not want associated with them in the real world.

His colleague Dave Willner is known as the "troll slayer." Officially, the twenty-seven-year-old who goes to work most days in blue jeans and a T-shirt develops policy for Facebook's "hate and harassment" team. These are the people responsible for enforcing a range of rules and policies meant to protect users from harassment and cyber-bullying. Both human and automated enforcement mechanisms aim to prevent the site from being overrun by spammers and criminals, which, he and his colleagues told me on a visit to Facebook headquarters, is an unending battle.

People want to be free to express themselves, organize whatever they want, and say whatever they want. Yet at the same time, parents want to keep their children safe from criminals, women do not want to be stalked by abusive ex-partners, and religious and ethnic rights groups will not tolerate the site's becoming a haven for hate speech or the launching pad of lynch mobs. The Simon Wiesenthal Center is unhappy that Facebook, on free speech grounds, refuses to shut down pages dedicated to denying that the Holocaust happened. But Facebook does shut down groups that cross the line from expressing opinions into more aggressive or organized campaigns against Jewish people. Many child protection organizations complain that Facebook has not done nearly enough to keep young people safe online. Such is the problem with governance, online and offline: going too far for some, not doing enough for others.

On any given week Facebook's "hate and harassment" team receives two million reports from users who have identified content they believe is abusive, harassing, or hateful and should be taken down. The problem is that the people who make abuse reports are not very "accurate." Only about 20 percent of these reports are for behavior or content that fit the definition of abusiveness according to Facebook's terms of service. Meanwhile, a lot of what the team would define as genuinely abusive never gets reported at all.

Thus a big part of the team's job is to develop processes to identify abusive content and remove it, while not removing other postings or pages that may be edgy and upsetting to some but are not actually against the terms of service. They have developed a system that combines automated software to identify image patterns, keywords, and communication patterns that tend to accompany abusive speech, along with review procedures by flesh-and-blood human staff. Willner focuses on defining policy for the site: guidelines about exactly what people should or shouldn't be allowed to do under what circumstances, and procedures for how violations are handled. These friendly and intelligent, young, blue jeans–wearing Californians play the roles of lawmakers, judge, jury, and police all at the same time. They operate a kind of private sovereignty in cyberspace.

"Overwhelmingly, most people won't engage in antisocial behavior if it's associated with their real-life identity," Willner told me. His bosses agree. In a 2004 interview with author David Kirkpatrick, Mark Zuckerberg alluded to a kind of social contract between Facebook and the user: "We always thought people would share more if we didn't let them do whatever they wanted," he said, "because it gave them some order." In 2010 Elliot Schrage, vice president for public policy, put it even more directly to the *Financial Times*: "We believe we are innovators in helping people manage their identities and reputations online, in contrast to the lack of control that exists on the Internet as a whole."

In seeking to build and maintain a global platform that can be used and trusted by hundreds of millions of people of all ages, cultures, and religions, Facebook has sought to shelter its users from a virtual version of what seventeenth-century British philosopher Thomas Hobbes famously called the "state of nature." In this primitive state, life is "nasty, brutish, and short," due to a complete lack of government. People are thus completely free to do whatever they like without any constraints or consequences. This state of complete freedom may be ideal for the strongest and most aggressive people, and maybe occasionally for the cleverest if they band together successfully, but not for most people most of the time. Rational individuals seeking to maximize their own self-interest will seek to replace the "state of nature" with some kind of government. Thus a "social contract" is formed: individuals recognize that it is in their interest to voluntarily relinquish a certain amount of their personal freedoms to a sovereign or government, which in turn sets and enforces rules meant to serve the interests of all members of society, in aggregate. Hobbes concluded that the greater good could be served only by a strong authoritarian sovereign with concentrated powers.

One of several major problems with Facebook's governance system, however, is that it is not enforced consistently or uniformly, and there has been no clear or straightforward appeals process for people who are not famous or do not have personal connections to members of Facebook's management team. In June 2010 a Facebook page with more than 800,000 members called "Boycott BP," created in response to the worst oil spill disaster in the continental United States, was disabled without warning for reasons that remain unclear to the group's creator, Desmond Perkins. After the takedown was reported on CNN's iReport and a variety of other news outlets, Facebook restored the page two weeks later. The explanation was terse: "The administrative profile of the BP Boycott page was disabled by our automated systems, therefore removing all the content that had been created by the profile. After a manual review, we determined the profile was removed in error, and it now has been restored along with the page." Greg Beck, an attorney for Public Citizen, a group supporting the BP boycott, told CNN in a follow-up story that he found Facebook's explanation frustrating. "Facebook and other social websites have become the public squares of the Internet—places where citizens can congregate as a community to share their opinions and voice their grievances," he said. "Facebook's ownership of this democratic forum carries great responsibility."

Many users—who like most people do not actually read the terms of service—are not even aware that using a fake name is against the rules. Inconsistent enforcement means that some people have gone for years using a fake name without a problem. A user-name search on Facebook for "Donald Duck" turns up many dozens of users by that name. The same thing happens with a search on the Chinese characters for "cola" (though Facebook is blocked in China, a lot of people in Hong Kong, Taiwan, and Singapore use it in Chinese). The abuse team says they cannot go after everybody and must prioritize the accounts that have unusual patterns of activity or that other users actively report as having violated the terms of service. This means that if a person is not using his or her real name and is engaging in controversial or high-profile activities on Facebook, that person is on particularly shaky ground—ground that Facebook reserves the right to pull out from beneath the user at any time. After all, the user consented to this situation when setting up the account by clicking "agree"—regardless of whether he or she read and understood the legal text being agreed to.

Critics are concerned that Facebook's core ideology—that all people should be transparent and public about their online identity and social relationships—is the product of a corporate culture based on the life experiences of relatively sheltered and affluent Americans who may be well intentioned but have never experienced genuine social, political, religious, or sexual vulnerability. As Microsoft researcher danah boyd (who officially spells her name in all lower case) wrote on her personal blog in the spring of 2010, "I think that it's high time that we take into consideration those whose lives aren't nearly as privileged as ours, those who aren't choosing to take the risks that we take, those who can't afford to." Activists from Iran are one example: their work would not be as effective if they could not use Facebook, but for them it is also too dangerous to use their real names.

Nobody is forcing anybody to use Facebook. Yet for political activists—or anyone trying to convince a large and diverse audience of anything—abandoning Facebook is easier said than done. In 2010, Americans spent more time on Facebook than on Google. If the largest pool of people your political or social movement most needs to reach is most easily and effectively reachable through Facebook's vast social network, leaving Facebook is a blow to the movement's overall impact.

It would be unfair to say that Facebook does not care at all about user opinion. It does in its own authoritarian sort of way, just as the Chinese government needs to care about public opinion if it wants to stay in power. In an online Q&A session on the *New York Times* "Bits" blog in May 2010, Facebook's public policy chief Elliot Schrage described how company staff had set up special pages and discussion groups inside the platform where people could comment on privacy policies, design features, and other policies. "Whenever we propose a change to any policies governing the site, we have notified users and solicited feedback," he said. Given the uproar over changes to Facebook privacy settings in late 2009 and early 2010, "clearly, this is not enough," he admitted. "We will soon ramp up our efforts to provide better guidance to those confused about how to control sharing and maintain privacy." Later in the session, in response to further critical questions by *New York Times* readers, he remarked: "It takes forums like this to get better ideas and insights about your needs."

Such comments sound eerily similar to those of Chinese Premier Wen Jiabao, congratulating his government in public web chats for caring so much about public opinion. Ultimately, however, just as China is governed by technocrats and functionaries who have no popular mandate other than a general claim that people's standard of living has improved dramatically over the past thirty years, decisions at Facebook are made by a group of managers who insist they are acting in users' best interests. Their belief is mainly supported by the fact that millions of people continue to join and use Facebook and that frequent Facebook users have been showing a trend of staying on the site for longer periods of time.

This assumption—that everything is fine as long as growth continues, and that the complainers are in the minority—is standard authoritarian fare. It reflects the classic Hobbesian social contract: a bargain between public and sovereign that Hobbes used to justify the need for enlightened monarchy. Hobbes was very much a royalist; Zuckerberg and company may have deployed the tools that people are using around the world in pushing for democracy but they are no democrats when it comes to balancing the rights and risks their users face.

In May 2010, a group of activists tried to get people to protest Facebook's power by deleting their accounts on a designated "Quit Facebook Day." Though 38,146 people pledged to quit, the effort made no meaningful dent in Facebook's growth from 400 million to 500 million users that year and seemed to have no impact on Facebook's policies. Reflecting on this failed boycott, danah boyd wrote an essay arguing that activism rather than boycott is likely to be more effective in bringing change to Facebookistan:

> Regardless of how the digerati feel about Facebook, millions of average people are deeply wedded to the site. They won't leave because the cost/benefit ratio is still in their favor. But that doesn't mean that they aren't suffering because of decisions being made about them and for them. What's at stake now is not whether or not Facebook will become passe, but whether or not Facebook will become evil. I think that we owe it to the users to challenge Facebook to live up to a higher standard, regardless of what we as individuals may gain or lose from their choices.

Activism by users, celebrity technologists, human rights organizations, and civil liberties groups has in fact managed to have some impact on the governance of Facebookistan. In response to sustained lobbying by the Electronic Frontier Foundation, the Committee to Protect Journalists, and others, Facebook's engineers added new encryption and security settings that enable users to better protect themselves against surveillance of as well as unauthorized intrusion into their accounts. After holding a number of conversations with activists and human rights groups, in mid-2011 the company rolled out an easy-to-use appeals process for people to contest the deactivation of their accounts, a process which until then had not existed for people without personal connections to Facebook staff. Facebook also developed a new "community standards" page to explain its terms of service in a simple and accessible way. Yet while that page as well as the official legally binding Terms of Service page were translated into Arabic, as of mid-2011 those key documents explaining Facebook's real-name policy and other "rules" whose violation could trigger account deactivation and suspension had yet to be translated into the languages of a number of other vulnerable user groups, such as Chinese and Farsi.

GOOGLE GOVERNANCE

After a year in self-imposed exile, Lokman Tsui decided that his departure had done nothing to help change Facebook and rejoined. Though his absence made no difference when it came to keeping up with close friends, he realized that he missed being able to easily stay in contact with the many other friends, former classmates, and colleagues with whom he otherwise had "weak ties." In a way, nobody else was punished by his exile but himself. Meanwhile, in June 2011 Tsui was getting ready to start a new job as a Hong Kong–based policy adviser for Google. In an e-mail explaining his decision to rejoin, he wrote, "I feel that I would do my new Google policy job a disservice by being disconnected from Facebook. That is, I need to be on there to know what is happening from a professional point of view." He had also come around to danah boyd's point of view: that engaging Facebook from the inside as a noisy constituent and customer might in the long run be more effective than flinging criticism from a position of exile. Ironically, Tsui would soon find himself on the other side of a firestorm over how Google governs its users.

In late June 2011, Google began a gradual rollout of its new social networking service, Google Plus. Despite being invitation-only for the first month or so, to give engineers and designers a chance to iron out the bugs, by mid-August it already had more than 20 million users. I joined on June 29 after receiving four invitations from friends who are all considered Internet gurus and experts of one kind or another. Thus the earliest users of Google Plus were for the most part experienced, web-savvy, and articulate people who immediately began to explore the network's features, discussing them in great detail and comparing them with Facebook's.

Initial reactions were largely positive. Many welcomed Google Plus's more sophisticated approach to privacy, giving users much more finely grained control over what they choose to share, with whom, and under what circumstances. A number of early flaws in the privacy control system were quickly fixed by engineers. Another feature called the data liberation front made it possible for users to extract all of their data if they decided to leave the service or wanted to back it up elsewhere for safekeeping. Such features were praised by civil liberties groups and activists, some of whom hoped that more competition would force all companies to improve their practices.

Many of Google Plus's earliest members also rejoiced at what seemed to be a more flexible approach to identity compared to Facebook's, citing its official "community standards" page, which said, "To help fight spam and prevent fake profiles, use the name your friends, family or co-workers usually call you." The Chinese blogger who publishes widely in English under the name Michael Anti—not actually his real Chinese name—happily joined Google Plus after having been kicked off Facebook for violating its real-name policy in late 2010. Many others around the world who have professional reputations associated with long-standing pseudonyms instead of their real names signed up for Google Plus with their pseudonyms. These included an

Iranian cyber-dissident known widely in the Iranian blogosphere as Vahid Online, as well as an engineer and former Google employee whose real name is Kirrily Robert but who is much better known online and professionally by her user name, Skud.

The honeymoon did not last long. In mid-July Google moved to deactivate pseudonymous accounts en masse, without warning. This came as a shock, particularly to many people whose pseudonyms are in fact the names by which they are commonly known "in daily life." Because a large number of these early Google Plus users happened to be bloggers, journalists, technologists, and activists, many protested their deactivations noisily in the blogosphere, in the media, and in their new Google Plus networks full of influential journalists and bloggers, who quickly relayed their stories to broader audiences. A UK-based science writer who publishes articles in the *Guardian* under the pseudonym GrrlScientist wrote an article about her experience. She quoted a Google spokesperson who explained that to have a Google Plus account, a person must have a Google profile, which in turn requires the use of one's real name, despite the fact that the legal right to use pseudonyms, even in very official contexts like financial transactions, paying taxes, and filing lawsuits, is well established in many countries. Effectively, Google had joined Facebook in denying people's right to define their own identity, a right that a large percentage of their users expect to be protected and respected even by government authorities.

As a former Google employee who had been privy to internal company discussions about identity policies prior to the launch of Google Plus, Skud had anticipated this situation and collected a vast number of web links and testimonials to prove that "Skud" is a persistent pseudonym tied to one individual who has taken responsibility for her actions and words over many years. After her Google Plus account predictably was deactivated, she included this evidence as part of her appeal for reinstatement via a formal appeals process Google had set up to handle mistaken deactivations.

Skud's appeal was denied and she eventually created a new account as "K Robert." But she was not ready to give up the larger battle to convince

Google to change its identity policy. She surveyed people in similar situations and found that although Google's appeals process encouraged but did not require people to upload copies of a government-issued ID to prove their identity, almost nobody had succeeded in having their account reinstated without uploading their ID. She launched the website my.nameis.me, dedicated to discussing questions of identity and why pseudonymity and anonymity have a necessary place in a free and democratic society. Technology gurus and activists from around the world weighed in, contributing statements and testimonials. Though few mainstream news organizations had written about the human rights implications of Facebook's real-name policy, the torrent of commentary flowing at the same time from many influential technologists—coming right on the heels of Google Plus's launch, which was in itself a major news story—brought the public debate about online identity into the mainstream in a way that had not previously been the case.

Then Randi Zuckerberg, Mark Zuckerberg's sister and Facebook's head of marketing, provoked a new round of controversy when she defended the need for real-ID rules. "I think anonymity on the Internet has to go away," she remarked at a conference. "People behave a lot better when they have their real names down. . . . I think people hide behind anonymity and they feel like they can say whatever they want behind closed doors." Activists countered that such attitudes leave no room for the Internet's most vulnerable users: from cyber-activists in Iran like Vahid Online (whose account was suspended in early August), to victims of domestic abuse, to people living in small communities where knowledge of their real sexual orientation or political views might incite reprisals or social ostracism of their families. Chinese Google Plus users began to noisily protest the policy in the service's help forum and elsewhere on the site, pointing out that even many Chinese social networking platforms allow pseudonyms.

One of the most sophisticated and original arguments in favor of pseudonymity on social networks like Facebook and Google Plus was articulated by Tunisian blogger and cyber-activist Slim Amamou in an e-mail exchange that included activists and Google employees. He argued that if the intent of a social network's identity policy is to create a trusting and safe environment, tying identity to cards issued by governments punishes the most vulnerable members of society by excluding them. In fact, when an online community includes many people who live in countries with repressive regimes, a company's insistence on real-name policies tied to government-issued ID actually erodes trust between users and the company, and even makes it harder for some users to trust one another. A trustworthy and responsible online identity, he explained, is "related to a public history, which is the definition of a profile in a social network. In other words *it's the profile who creates identity* through trust and not the other way around. I repeat, it's you, Google Plus, who are supposed to generate identities and not simply trust nation states' administrations for that."

Such arguments, however, were unsuccessful in budging Google's top management. In early August, the company reaffirmed its real-ID policy for Google Plus but instituted a new four-day grace period between when users are warned that they are in violation and the deactivation of their account. Privately and off the record, a number of Google employees told me in July and August 2011 that there had been fierce internal debate about how identity should be handled on Google Plus. In the end, the decision was made at the highest levels of the company that in order to make Google Plus a commercially successful platform, real-name identity would need to be enforced. Google Plus simply was not created with dissidents in mind and was not meant to be used for political dissent or by other people who are not comfortable disclosing their real identities. Other Google services including Blogger, YouTube, and Gmail would continue to support pseudonyms, and the company would remain dedicated to protecting the rights of dissidents to use those services.

This was not enough for people who had hoped that Google Plus would be a dissident-friendly Facebook alternative. Many people continued to protest, some deliberately setting up pseudonymous accounts to publicize their deactivation; others led long tactical and strategic discussions about how to engage Google management and convince them to change their policy. These people's perseverance is important. As social networks become an increasingly influential part of citizens' political lives, telling political dissidents, human rights activists, and other at-risk individuals that there is no place for them on the world's most popular and widely used social networks has serious political implications on a global scale. The harm is no less real even when the companies that run those networks genuinely do not mean to cause harm.

IMPLICATIONS

In the long run, if social networking services are going to be compatible with democracy, activism, and human rights, their approach to governance must evolve. Right now, for all their many differences, both Google Plus and Facebook share a Hobbesian approach to governance in which people agree to relinquish a certain amount of freedom to a benevolent sovereign who in turn provides security and other services.

Fortunately, Hobbes was by no means the last word in social contract theory. He was followed by John Locke, one of the first political thinkers to set forth a logical argument for why government should be based on "consent of the governed," the fundamental idea that inspired the English, American, French, and other more recent revolutions. In Locke's *Second Treatise of Government*, a document to which Thomas Jefferson turned in drafting the Declaration of Independence, government is legitimate only when it satisfies the fundamental needs of the community. A government that violates the trust of its people loses their "consent"—and therefore deserves to be overthrown.

Locke drew inspiration from a rebellious group of men known as "The Levellers," an informal alliance of agitators and pamphleteers during the English civil war of the 1640s who believed the monarchy should be abolished and replaced by a civil state based on English common law and a few statutes including the Magna Carta, which over four centuries earlier represented the first effort to place constraints on sovereign power.

The modern sovereign—otherwise known as government—derives authority even to some extent beyond the community of parliamentary democracies, from varying forms and degrees of consent. It is time for the new digital sovereigns to recognize that their own legitimacy—their social if not legal license to operate—depends on whether they too will sufficiently respect citizens' rights.

The social contract on which modern democracy is based is primarily concerned with the protection and respect for citizens' property and physical liberty. As citizens, we now use digital networks and platforms—including Facebook and now Google Plus—to defend our physical rights against abuse by whatever physical sovereign power we happen to live under, and to bring about political change. However, our ability to use these platforms effectively depends on several key factors that are controlled most directly by the new digital sovereigns: they control who knows what about our identities under what circumstances; our access to information; our ability to transmit and share information publicly and privately; and even whom and what we can know. How our digital sovereigns exert these new powers may or may not be in response to government laws or pressures, and those pressures may be direct or indirect. Either way, the companies controlling our digital networks and platforms represent pivotal points of control over our relationship with the rest of society and with government.

No company will ever be perfect—just as no sovereign will ever be perfect no matter how well intentioned and virtuous a king, queen, or benevolent dictator might be. But that is the point: right now our social contract with the digital sovereigns is at a primitive, Hobbesian, royalist level. If we are lucky we get a good sovereign, and we pray that his son or chosen successor is not evil. There is a reason most people no longer accept that sort of sovereignty. It is time to upgrade the social contract over the governance of our digital lives to a Lockean level, so that the management of our identities and our access to information can more genuinely and sincerely reflect the consent of the networked.

PART FIVE

WHAT IS TO
BE DONE?

Sir, I think it's clear that every man that
is to live under a Government ought first
by his own consent to put himself under
that Government.

—"LEVELLER" COLONEL THOMAS
RAINSBOROUGH, debating the English
royalists at Putney Church, 1647

CHAPTER  11

Trust, but Verify

In November 2010, I sat in on a meeting of the International Engineering Task Force, which coordinates technical standards for the technologies that make the Internet functional and globally interoperable. I listened as several engineers employed by Cisco argued passionately in favor of technical standards that can help to protect the privacy and anonymity of individual Internet users, sometimes in opposition to engineers from China and Russia whose employers clearly have other priorities. Such efforts by a number of senior Cisco engineers in highly technical forums whose work is not usually covered by journalists receive little attention beyond a small technical community. In June 2011, Senior Vice President and General Counsel Mark Chandler wrote on the company's blog, "Our goal in providing networking technology is to expand the reach of communications systems, and our products are built on open, global standards. We do not support attempts by governments to balkanize the Internet or create a 'closed' Internet because such attempts undermine the cause of freedom. In fact, adherence to open standards is critical in the efforts to overcome censorship." Such statements, which do in fact appear to be backed up by Cisco's actions, have been overshadowed by other controversial aspects of Cisco's global business, particularly in China.

What everybody who follows or covers news about the Chinese Internet *does* know is that Cisco has been in the doghouse with human rights groups and some socially responsible investors for nearly a decade, because its routers are used in China as part of the "great firewall" blocking system

169

and because it does business with Chinese police departments. Cisco executives argue that their role in China and elsewhere has been misunderstood: at a 2006 congressional hearing during which several companies were called on the carpet about their activities in China, Chandler argued that his company sells the same equipment worldwide and "does not customize, or develop specialized or unique filtering capabilities, in order to enable different regimes to block access to information."

In July 2011 the *Wall Street Journal* reported that Cisco, along with other Western companies, would supply some of the networking equipment to the city of Chongqing, in southwestern China, for one of the "the largest and most sophisticated video-surveillance projects of its kind in China," linking up as many as 500,000 surveillance cameras and covering an area 25 percent larger than New York City. The report did quote a Cisco spokesman who emphasized that the company "hasn't sold video cameras or video-surveillance solutions in any of our public infrastructure projects in China." In a blog post responding to the article, Chandler wrote, "We were offered an opportunity to supply those products in Chongqing and, contrary to the suggestion in the article, declined that opportunity." He emphasized that Cisco's proposed participation in a bid to sell "standard, unmodified Cisco routing and switching equipment— the same equipment that is supplied to governments and private sector customers worldwide," is part of a broader initiative by the city of Chongqing called "Smart+Connected Communities," which includes e-government services, "smart building" technology for energy efficiency, and the improved sharing of educational and health-care information.

Even if Chandler's explanation is true, however, it is also clear that Cisco has sought business opportunities with law enforcement authorities in China, a country that clearly defines "crime" to include political and religious dissent. In 2005 the author Ethan Gutmann published Chinese-language marketing brochures for surveillance equipment, distributed by Cisco at a law enforcement trade fair he had attended in 2002. In 2008 activists published a PowerPoint presentation by a Cisco marketing manager, citing the Chinese government's campaign against the banned Falun Gong religious group as one of many reasons there is demand for Cisco's products by Chinese security and law enforcement organs. These incidents raised questions from human rights groups and several institutional investors about the extent to which Cisco may have actively assisted or even promoted their technology to customers who clearly intended to use it for repressive purposes.

Yet Cisco refused to hold a substantive conversation about these questions even with its own shareholders. After five years of failed efforts to convince Cisco to engage with shareholders in a dialogue about the marketing and application of its technologies in ways that might help the company avoid complicity in human rights violations, the socially responsible investment company Boston Common Asset Management divested all of its Cisco stock in January 2011. Other socially responsible investment firms like Domini and Calvert continue to press the company to at least acknowledge its specific human rights risks and responsibilities it shares and address the real dilemmas it faces—a simple step that has seemed difficult for the company to take.

To be fair, there are many other companies doing business with police departments in China and in other countries whose definition of "crime" often includes political and religious dissent. Most of these companies have received much less attention than Cisco has. Take, for example, Hewlett-Packard, which according to the *Wall Street Journal* also planned to bid on the Chongqing project. Asked by a reporter to explain HP's participation, Todd Bradley, the executive vice president in charge of the company's China strategy, responded, "We take them at their word as to the usage. . . . It's not my job to really understand what they're going to use it for. Our job is to respond to the bid that they've made." HP understands that it has an obligation to ensure that workers are not abused in the course of manufacturing its product and that its products must be manufactured in an environmentally responsible way, or it will be punished in the long run by consumers, investors, and regulators. Clearly, however, the company does not seem to think it should be held responsible in situations where its products have a good chance of contributing to repression.

Unlike companies that produce sportswear or toothpaste, the value proposition of Internet-related companies relates directly to the empowerment of citizens. Platforms like Twitter and Facebook have won the fawning adoration of media, policy makers, and much of the general public because they have made possible positive, historic change in the world by creating whole digital worlds through which citizens can congregate in cyberspace and share information, ideas, and offline action. Yet at the same time, most of these companies have demonstrated a shocking blind spot when it comes to a key element of public trust: accountability. Having altruistic-sounding mission statements on the corporate website is well and good, but how can people be sure a company is living up to its own ostensibly high ethical standards—any more than they should trust that a sovereign is good simply because he says he is? In the long run, an Internet-related company's value proposition is questionable at best and fraudulent at worst if it rejects the need for accountability.

As Ronald Reagan famously said to Mikhail Gorbachev when they signed a major arms control treaty in 1987, "Trust, but verify." As citizens, we are right to hold the same attitude toward Internet and mobile communication companies, which we now depend upon to inform ourselves, participate in political discourse, and exercise our rights as citizens. We need to be able to trust these companies upon whose platforms, services, and technologies we increasingly rely. Taking them at their word and hoping for the best is no more likely to work out well for us than when we take government unquestioningly at its word.

In many situations, government regulation of companies to compel the protection of citizen rights is essential, particularly when there is a substantial body of evidence that the companies in question may not be willing or able to protect citizen rights of their own accord. There is a need for regulation and legislation based on solid data and research (as opposed to whatever gets handed to legislative staffers by lobbyists) as well as consultation with a genuinely broad cross-section of people and groups affected by the problem the legislation seeks to solve, along with those likely to be affected by the proposed solutions. In many other situations, government regulation—especially when large numbers of people have good reason not to trust the motives of the regulators or legislators in question—can create as many problems as it solves.

The world's netizens will remain impaled on the horns of this dilemma without progress on at least two fronts: First, we must come up with more innovative ways to hold companies accountable for the human rights implications of their business, software, and engineering choices. Second, people and organizations concerned about the problems described in this book need to work more creatively and aggressively to help companies shape their products and services in a way that will minimize the likelihood that their businesses will violate—or facilitate the violation of—human rights and civil liberties, and maximize the chances that their businesses will genuinely improve the world.

THE REGULATION PROBLEM

In most countries the primary way of holding companies accountable to the public interest is by passing laws. Politicians in many of the world's democracies are calling for tough privacy laws that would manage how companies can collect and use people's information. Regulation is clearly needed. Many companies, however, are concerned that the provisions in many countries' privacy laws fail to take into account the realities of technical innovation and business models, are excessively intrusive into corporate processes, and will stifle innovation. They sometimes have a point, depending on the specifics of the law and the particular issue it seeks to address. But if enough people feel they cannot trust Internet and telecommunications companies to be honest about what data is gathered about users and customers—with whom and how it is shared, and why—the companies cannot reasonably expect to not be regulated with increasing aggressiveness. As for companies concerned that net neutrality legislation will stifle innovation, constrain commerce, and inhibit companies' ability to provide new and useful services to the public, the onus is similarly on the companies to prove they can be trusted without being regulated. So far, too many large companies have simply provided the public with too many compelling reasons not to trust them.

To address corporate complicity with authoritarian censorship and surveillance, for the past decade some members of Congress and many human rights groups have supported a bill called the Global Online Freedom Act (GOFA). One of GOFA's many problems, however, is that when it was originally written in 2004, it sought to address the problems faced by and created by companies operating in China at the time—which in Internet time is ancient history. The problems have long since evolved, spread much more globally, and grown much more complex. Furthermore, because every company's technology, business model, and business relationships around the world are so different, it has proven impossible to adopt an effective one-size-fits-all legislative approach without becoming so broad as to be meaningless—or so restrictive as to make it impossible for US technology companies to do business in many countries (most obviously but not exclusively China). Blocking US Internet and telecommunications companies from ever operating in authoritarian or quasi-democratic countries amounts to counterproductive overkill, preventing citizens from using some of the world's most innovative and open technology to advocate for change, let alone shutting companies out of massive markets that are no longer "emerging" but are now well established.

A broader and more intractable problem with regulating technology companies is that legislation appears much too late in corporate innovation and business cycles. Legislation does nothing to help companies anticipate problems that their business or design choices might create—before they implement them. By the time a law is called for, the damage has already been done and alternative paths have long been closed off.

A further problem with GOFA, even after congressional staffers attempted to revise and update it, was its focus on regulating business activities only in countries the State Department designated as "Internet-restricting countries." Given the expansion of censorship and surveillance throughout many democratic countries, however, where does one draw the line? What happens in cases where, for example, a country was democratic last week, then experienced a coup and suddenly became a military dictatorship overnight? Since that country would not be on the State Department's list, companies would not be held legally accountable for behavior there.

Furthermore, there is no country on earth whose government has not in some way sought to extend its power through private networks in ways that many people believe to be an abuse of power. The difference, of course, is the extent to which the government in question can be held accountable, its power appropriately constrained through legal and political institutions and processes. The US government has itself contributed to the erosion of due process and accountability in its relationship with privately owned and operated networks, expanding surveillance powers and making it legally easier to take down websites, insisting that this is a necessary trade-off to fight crime, terrorism, and cyber-attacks. It therefore seems difficult to expect that any government—no matter how democratic or well intentioned it claims to be—can be fully trusted to prevent the violation of citizens' rights caused by companies' unaccountable and opaque compliance with government demands, or due to the sale of certain technologies to governments.

Because government is often part of the problem even in democracies, corporate collusion in government surveillance and censorship is unlikely to be solved by the passage and enforcement of laws, even by the most well-intentioned and democratic of governments. Companies must also be convinced that respecting and protecting their users' universally recognized human rights is in their long-term commercial self-interest—a proposition that continues to puzzle or elude too many Internet companies, in contrast to longer established technology companies and others in virtually every other industry.

SHARED VALUE

When it comes to the physical environment, corporate responsibility and sustainable business are now clearly related, widely accepted, and irrevocably mainstream concepts. Thanks to decades of activism by environmental, human rights, and consumer groups, as well as socially responsible investors, it is now normal for consumers, mainstream investors, and legislators to expect that businesses will operate in ways that respect the environment, their workforces, and the quality of life and basic human rights of the communities in which they operate.

In the short run, companies that pollute the groundwater and expose workers to toxic conditions may maximize profits, generate jobs, and contribute to a country's economic growth. In the long run, we know that such unsustainable practices destroy the environment, threaten public health, and impose serious long-term costs on society and the planet. Business has a long-term incentive to operate more sustainably: a company that contributes to a toxic environment and an unhealthy society might make its shareholders richer in the short run, but in the long run it cannot thrive at the expense of its customers, its employees, and the communities and countries where it operates.

Millions of companies worldwide have embraced this notion to varying degrees by building corporate social responsibility (CSR) and sustainability strategies into the way they run their businesses. A growing number of investors are rewarding companies for moving in this direction. By 2010 the amount of assets held by US investors in some form of socially responsible investment funds, which choose stocks based on at least some criteria for sustainable and socially responsible business practices, had reached $3.07 trillion out of a total of $25.2 trillion in the US investment marketplace. A range of UN agencies, financial institutions, and nongovernmental groups have developed systems to benchmark companies based on various criteria for corporate social responsibility or sustainability.

Meanwhile, a growing number of multinationals are moving beyond CSR programs to identify untapped business opportunities around growing public demand for cleaner, healthier, and more sustainable lifestyles in ways that can become a source of competitive advantage and improve the world at the same time. Innovation to create new products, services, manufacturing processes, and business practices that directly contribute to improved human health and environmental sustainability is part of what Harvard Business School gurus Michael Porter and Mark Kramer call "shared value." In a January 2011 *Harvard Business Review* article, they point to a growing body of research that demonstrates when a business brings its operations and business strategies into direct alignment with society's broader aspirations and values, it maximizes its own ability to "drive the next wave of innovation and productivity growth in the global economy."

A handful of socially responsible investment funds, including Calvert, Domini, and Boston Common in the United States, as well as F&C Asset Management and Folksam in Europe, now include digital free expression and privacy as part of their investment criteria. So far, however, most Internet-related companies have failed to apply the concept of public trust, sustainability, or shared value to the digital public spheres they are responsible for creating, shaping, and governing. They have failed to address how their actions affect the Internet's long-term viability and credibility as an open and globally interconnected network. Or more disturbingly, they do understand but do not care because the majority of their users, customers, and investors continue to reward them even as they engage in practices—and in some cases even sell technologies—that make the Internet less open and free.

When Google decided to stop censoring its Chinese search engine, Google.cn, and moved it from mainland China to Hong Kong, many commentators noted cynically that Google's move was motivated as much by self-interest as it was by the founders' strong concerns about free speech. Google, they pointed out, had lost out to its domestic competitor Baidu when it came to market share. What most failed to understand, however, was that Google's decision was neither simply about the company's business—or lack thereof—in the Chinese market, nor was it only about the egos of the company's founders or their personal value systems (although those factors certainly played into the decision). It was also a statement about Google's relationships with governments and citizens worldwide, and about the entire Internet's future. Google made a decision that was both good for Internet freedom *and* in its long-term self-interest.

Though Google has many flaws and makes its share of mistakes, it nonetheless deserves praise for aligning its long-term interest with a public commitment to keeping the Internet open and free, even when doing so creates government relations nightmares—and loss of business in countries such as China—for some years. Standing up against government censorship and surveillance requirements that are in clear violation of international human rights norms is an investment in the Internet's long-term sustainability and long-term value. By operating a censored search engine in China, Google was lending credibility and legitimacy to the Chinese government's model of networked authoritarianism. If more governments emulate the Chinese model, the value of the Internet as a whole—and the value of services that global Internet companies seek to provide—will be greatly corroded. People's loss of trust in those who control and shape their digital lives will also erode the value of Internet-related products and services over the long term. The Internet's value will therefore be much less sustainable if most companies' default mode is to accommodate the demands of authoritarian regimes—and overreaching governments more generally—for the sake of short- to medium-term market access and profit.

If one examines Google's disputes with authoritarian regimes as well as democracies around the time of its withdrawal from China, a picture emerges. In late 2009, several months before Senior Vice President and Chief Legal Officer David Drummond first published his company's game-changing decision about its China operations on the company's official blog, the site was full of postings by senior executives about the company's conflicts with a range of governments. In one December 2009 posting, Public Policy Director Bob Boorstin blasted the Australian government's censorship efforts. Another executive posted an item about an anticensorship workshop held at Google's Mountain View, California, headquarters, in which the world's top technical experts and academic researchers who study censorship and how to get around it spent a day brainstorming with Google employees about combating the spread of Internet censorship all over the world. These postings were followed by statements explaining Google's fights with the French and Italian governments, along with discussion of the ACTA trade agreement on copyright enforcement—supported enthusiastically by the US entertainment industry—which Google lobbied strongly against.

In the midst of all these blog postings, one that best explains the global context of Google's China decision is a long memo titled "The Meaning of Open" posted on December 21, 2009, by Jonathan Rosenberg, senior vice president for product management. He wrote:

Closed systems are well-defined and profitable, but only for those who control them. Open systems are chaotic and profitable, but only for those who understand them well and move faster than everyone else. Closed systems grow quickly while open systems evolve more slowly, so placing your bets on open requires the optimism, will, and means to think long term.

Google is betting its future on an open and largely free Internet—which means that its decisions and actions everywhere in the world need to be consistent with that bet.

That said, Google continues to commit errors in earning public trust, particularly when it comes to questions of how information on its users' online activities are stored and used, and more recently in its handling of identity policies for its new social network, Google Plus. Just because Google has done the right thing on several fronts certainly does not mean the company will on others, or that people should automatically trust that its executives, programmers, and designers are inherently more likely to make the right decisions, or will even have the right intentions in all situations. We would be foolish to trust the sovereigns of our physical world in that way. There is no good reason to be any more trusting of our digital sovereigns.

THE GLOBAL NETWORK INITIATIVE

Recognition of this reality—largely forced by congressional threat of legislation in 2006—is what brought Yahoo, Google, and Microsoft to the bargaining table with human rights groups, socially responsible investors, and academics to set up a multi-stakeholder process focused on free expression and privacy. Following nearly three years of dialogue and negotiation, in late 2008 they launched the Global Network Initiative (GNI), dedicated to helping Internet and telecommunications companies uphold their users' and customers' rights to freedom of expression and privacy around the world in ways that are credible and accountable. (Disclosure: I helped to launch the organization and am currently on the GNI's board of directors.)

The GNI's key challenge is this: Given that there is basically no country on earth where government is not pressuring companies to do things that arguably infringe on citizens' rights, how do companies take practical steps to protect their customers' and users' rights to free expression and privacy? The potential answers are daunting in their complexity: Every country's legal and political system is different. Almost every Internet platform, equipment vendor, and telecommunications company is different in terms of its technologies and business models. Every company conducts business differently in every country in which it operates.

Because of this complexity and diversity, creating a specific set of rules for all companies operating in all countries is a nonstarter. The whole point is to minimize harm while maximizing the good that the spread of information technologies brings to people around the world. If companies are hamstrung by rules that prevent them from doing business at all in much of the world, due to the propensity of so many governments to censor and monitor people in ways that arguably infringe on civil liberties, that result is clearly in the interest of neither users nor the broader public.

Companies that join the GNI first pledge to uphold a set of core principles. On freedom of expression, this includes a commitment to "respect and protect" the rights of users and customers "when confronted with government demands, laws and regulations to suppress freedom of expression, remove content or otherwise limit access to information and ideas in a manner inconsistent with internationally recognized laws and standards." On privacy, companies agree, among other things, "to respect and protect the privacy rights of users when confronted with government demands, laws or regulations that compromise privacy in a manner inconsistent with internationally recognized laws and standards."

The GNI's membership includes other stakeholders, such as human rights organizations, socially responsible investors, and academic specialists whose research and knowledge can help the companies both identify problems and explore solutions to them. Over the course of three years leading up to the GNI's launch, people representing the companies and other groups negotiated a set of operational guidelines, including the commitment by companies to conduct "human rights risk assessments" before introducing a new product or going into a new market.

One example of how such risk assessment works was Yahoo's handling of its Vietnamese business. When the Internet took off in Vietnam, Yahoo 360 (a now largely defunct blogging and social networking platform) happened to be where the majority of Vietnamese-language bloggers chose to congregate. Seeing a business opportunity, Yahoo explored setting up a local Vietnamese-language service to serve Yahoo's Vietnamese users more effectively. After conducting a human rights assessment, however, company executives decided that Vietnam's record of arresting political bloggers made it likely that if Yahoo began hosting its Vietnamese-language service in-country, its local staff would be forced to turn in local bloggers to the police just as Chinese staff had been forced to turn in Shi Tao's e-mail information. In the end, Yahoo decided to set up its Vietnamese-language operations in neighboring Singapore, where Vietnamese user data would be of little interest to Singaporean authorities, and out of reach of Vietnamese police.

There is also a rigorous external assessment process—call it a "human rights audit"—that aims to determine the extent to which companies are actually upholding their commitments and to identify areas that need improvement. GNI members will then work with companies to figure out how to make these improvements. This process aims to make sure that a company's membership in the GNI is not an empty public relations ploy.

Internal and external assessments cannot possibly anticipate or prevent every possible problem, which is why another important function of the GNI is to build a trusted channel through which companies can seek advice from human rights groups and academics when trying to resolve problems before they escalate. One example is when the Russian government targeted human rights activists and journalists on the pretext that they were using pirated Microsoft software. Microsoft worked closely with human rights groups in the GNI to identify the root of the problem, address it publicly, and find solutions that would prevent antipiracy campaigns from being used as a ruse to arrest human rights activists.

The first round of independent assessments for the three founding companies is scheduled for completion in early 2012. The GNI's efficacy in helping those three companies avoid business practices that violate citizens' rights is yet to be publicly demonstrated. But even if these three companies prove that their participation in the GNI reflects their accountability and reinforces user and public trust, the initiative will fail to achieve its potential as a global standard embraced by a critical mass of companies unless others see that it is in their interest to join. As of late 2011, no new companies had joined the GNI beyond the three founding members, Google, Microsoft, and Yahoo—despite the fact that Nokia Siemens, Ericsson, and Vodafone have all come under fire for assisting government suppression in Iran, Belarus, and Egypt, respectively. Facebook and Twitter have declined to respond when asked by journalists why they have not joined.

LESSONS FROM OTHER INDUSTRIES

Most Internet-related companies—in love with the popular media mythology that they are changing the world for the better just by existing—apparently do not understand why it is good for their business in the long run to be both responsible and publicly accountable when it comes to protecting users' and customers' rights. They fail to recognize how a company that exercises power without accountability, all the while making populist claims about empowering the masses (even if these claims are in many ways true), is ultimately offering nothing more than a corporate variant of digital bonapartism.

Meanwhile, many old-fashioned brick-and-mortar companies in the food, beverage, and fashion industries, and even a number of oil, gas, and mining companies, are much further along than the world's most cutting-edge and disruptive Internet-related companies when it comes to ac-countability. The GNI's framework for holding companies accountable, and working proactively to anticipate problems, was not dreamed up out of thin air. It was modeled after a number of existing multi-stakeholder initiatives, several of which have been operating for a decade or more, as part of an effort to hold companies accountable to basic environmental, labor, and human rights standards when governments fail to do so. The point is not to turn companies into social welfare or human rights organ-izations, but to help (and when necessary, prod) companies to achieve the maximum overlap between doing well and doing good.

The Fair Labor Association (FLA), established in 1999, is a platform through which companies work with NGOs and university purchasers of branded apparel and footwear to address human rights problems in these industries by implementing an agreed code of conduct. The association was set up because domestic and international laws on their own have proven inadequate when it comes to preventing corporate abuses of work-ers' rights. Like that of the GNI, the FLA's scope includes engagement by activists with companies, academic research, and information-sharing about best practices, and a framework for company assessment and ac-creditation. The idea is not only to provide data and benchmarks that stu-dents, universities, socially responsible investors, and customers of all stripes, as well as policy makers, can use to make decisions. The point is also to work positively and constructively with companies on manage-ment and factory innovations that will improve working conditions and ensure that inevitable violations will be reported.

The extractives sector has given rise to the greatest number of long-standing multi-stakeholder initiatives. The Voluntary Principles on Se-curity and Human Rights, launched in 2000, is a set of guidelines for oil, gas, and mining companies—which often operate in conflict zones and countries where the military and police act with impunity—to balance their necessary security arrangements with human rights safeguards to prevent complicity in violence against civilians. The Kimberley Process, established in 2003, is a certification scheme for diamonds and part of an effort to stem the flow of "conflict diamonds" from war-torn parts of the world. The Extractive Industries Transparency Initiative, launched in 2003, is another multi-stakeholder initiative through which more than fifty oil, gas, and mining companies from around the world have committed to full public disclosure and verification of revenue payments made to governments in the countries in which they operate. The Voluntary Principles and Kimberley Process in particular have faced their own challenges over meeting human rights groups' expectations about accountability. Yet they embody serious commitments by industries that are hardly seen as progressive or innovative by the sovereigns of cyberspace—but which have made more serious commitments than most Internet companies have been willing to make.

Closer to home, as far as Silicon Valley is concerned, for more than a decade hardware manufacturers like Hewlett-Packard have addressed labor issues and workplace standards in suppliers and factories, both on their own and through industrywide initiatives, such as the Electronic Industry Code of Conduct. More recently in 2009 and 2010, HP and AMD have taken leading roles, together with Ford and GE, in supporting legislation that requires certification for electronic components, in an effort to address the problem of "conflict minerals" in the war-torn Democratic Republic of Congo.

One would expect that if footwear and apparel manufacturers like Nike and Adidas can implement the FLA code of conduct in supplier factories, and that if oil companies like BP and ExxonMobil can implement the Voluntary Principles on Security and Human Rights in conflict zones, then Facebook and RIM can and should at the very least commit to the reasonable freedom of expression and privacy standards set forth by GNI. The sovereigns of cyberspace have no monopoly on virtue, and their empowering technologies are behind many "old economy" industries and companies, as well as computer and telecommunications hardware manufacturers, when it comes to certain aspects of corporate responsibility.

None of these initiatives comes close to preventing all abuses. In an ideal world, forward-thinking and well-crafted laws, passed by democratically elected legislatures and enforced by highly effective, publicly accountable government agencies, would be sufficient to encourage good corporate behavior and to discourage bad behavior. The real world, unfortunately, falls terribly short of that ideal; many companies find themselves operating in countries where authoritarian governments expect them to commit or contribute to human rights abuses—situations that make the challenges that oil and mining companies face in coping with security forces protecting their pipelines in conflict zones most analogous to what Internet companies face with censorship and surveillance. Unless and until the problem of incompetent, abusive, and unaccountable government is solved everywhere in the world, a range of multi-stakeholder efforts, which vary in their scope and composition from industry to industry, will remain necessary for a long time to come.

Industry, civil society, and some governments have now been working together for over a decade to establish various ad hoc frameworks through which companies can recognize and uphold their human rights obligations—while still operating as businesses. The challenge is to distill the lessons learned, so that citizens seeking to hold companies accountable are not forced to reinvent the wheel each time, with each new industry and sector (as we are now doing with Internet-related companies). To that end, in June 2011 the UN Human Rights Council approved the Guiding Principles on Business and Human Rights, the result of six years of research and consultation with companies, governments, and human rights groups by John Ruggie, the UN's special representative on business and human rights. The Ruggie principles outline three concrete steps that companies must take: first, make a "policy commitment to meet their responsibility to protect human rights"; second, develop a "human rights due-diligence process to identify, prevent, mitigate and account for how they address their impacts on human rights"; and third, initiate "processes to enable the remediation of any adverse human rights impacts they cause or to which they contribute."

Companies seeking to implement all three of these steps prove to their users, customers, investors, and regulators that they can be trusted not to abuse and violate the public interest. Multi-stakeholder initiatives like the GNI provide a source of human rights expertise, academic research, and independent verification, all of which bring substance and credibility to a company's public claims that it is making serious efforts to uphold human rights—and not just engaging in a fancy yet cynical public relations exercise. They work from the premise that companies face not only human rights risks and responsibilities but also dilemmas—and that no company will ever be perfect but that all companies will be held accountable.

The GNI's multi-stakeholder approach is the first concrete attempt by Internet-related companies to build public credibility, trust, and legitimacy—by making specific commitments, then undertaking a process to prove and independently verify what they have done to meet those commitments. Some human rights groups and Internet freedom activists have criticized the GNI for not having sufficiently broad international membership among noncompany members, for being too narrow in scope, and for setting the bar too low for companies. They certainly have a point. Yet there are currently no other functioning alternatives. The GNI is the only coherent effort so far by which any Internet-related companies have agreed to be held publicly accountable in any meaningful way for the impact of their businesses on human rights around the world. Companies in the GNI must be held to account for their commitments—and so too must the far greater number of companies that share similar risks and responsibilities explain just how they are addressing them, if not through the GNI. If most Internet-related companies cannot even step over what many people in the human rights community consider to be a low bar, that does not bode well for the future of human rights and civil liberties in the Internet age.

Meanwhile, governments of the world's democracies have only just begun to pay attention to the Internet freedom issue in any kind of systematic way and are in the early stages of formulating policies. To put it politely, participants in these policy processes and debates have faced a steep learning curve and lack of consensus about objectives, as well as muddled thinking when it comes to defining those objectives.

CHAPTER 12

In Search of
"Internet Freedom" Policy

On January 21, 2010, Secretary of State Hillary Clinton strode into the Newseum—a glass-and-steel shrine to American-style free speech just down Pennsylvania Avenue from Capitol Hill—and gave an impassioned forty-five-minute speech declaring "Internet freedom" to be a new pillar of American foreign policy. "On their own, new technologies do not take sides in the struggle for freedom and progress, but the United States does," she said. "Our responsibility to help ensure the free exchange of ideas goes back to the birth of our republic."

Just a few months earlier, Clinton had joined other world leaders atop Berlin's Brandenburg Gate to celebrate the twentieth anniversary of the fall of European communism. The parallels between the high-tech "firewalls" of censorship and the Soviet Iron Curtain were irresistible. "A new information curtain is descending across much of the world," she told the audience at the Newseum, channeling Winston Churchill's famous 1946 speech in which he warned that an "Iron Curtain" was descending across Europe. "And beyond this partition, viral videos and blog posts are becoming the samizdat of our day."

With Clinton's speech, "Internet freedom" became a prominent component of US diplomacy, alongside human rights and democracy-promotion policies. The problem, however, is that the phrase "Internet freedom" is like a Rorschach inkblot test: different people look at the same ink splotch and see very different things.

"Internet freedom" has many possible meanings. It can mean freedom *through* the Internet: the use of the Internet by citizens to achieve freedom from political oppression. It can mean freedom *for* the Internet: noninterference in the Internet's networks and platforms by governments or other entities. It can mean freedom *within* the Internet: individuals speaking and interacting in this virtual space have the same right to virtual free expression and assembly as they have to the physical pre-Internet equivalents. It can mean freedom *to connect* to the Internet: any attempt to prevent citizens from accessing it is a violation of their right to free expression and assembly. Finally, "Internet freedom" can also mean freedom *of* the Internet: free and open architecture and governance, which means that the people and organizations who use computer code to determine its technical standards, as well as those who use legal code to regulate what can and cannot be done within and through the Internet, all share the common goal of keeping the Internet open, free, and globally interconnected so that all netizens are free not only to use it, but also to participate in shaping it themselves.

WASHINGTON SQUABBLES

Global Internet freedom policy in Washington began with an emphasis on the first category: freedom *through* the Internet, a notion sometimes also referred to as "liberation technology." Under the previous administration of President George W. Bush, the State Department had already begun a program to support technologies for circumventing censorship. Between 2007 and 2010 Congress "earmarked" a total of $50 million to support technologies that would "tear down the firewalls" in countries where Internet censorship is heaviest, particularly China and Iran. Clinton made it clear that the State Department's initiative to support global Internet freedom would be broadened.

After her 2010 speech, a battle ensued not only over how the earmarked funds should be spent, but also over the definition of "Internet freedom," and the purpose of such a policy. One set of people in Washington understand "Internet freedom" to mean that the United States needs to help political and religious activists in undemocratic countries access an uncensored Internet and evade surveillance; thus their proposed solutions revolve around the funding of anticensorship and antisurveillance technologies. Another set of people view "Internet freedom" in a more global and universal sense, to mean keeping the Internet's architecture free and open everywhere in the world.

A subset of the first camp, which lobbied hard for Congress to allocate funds for circumvention technologies, was the Global Internet Freedom Consortium (GIFC), run by practitioners of the Falun Gong, a religious sect banned in China. The GIFC produced a suite of circumvention tools (Freegate, Dynaweb, Ultrasurf, and Ultrareach) that are effective at bypassing Internet blocking, as long as the user does not mind that Falun Gong–affiliated engineers can view and log their unencrypted communications or that the software's security—and thus its vulnerability to attack, infiltration, and data theft—has not been audited by independent experts. Other groups that build circumvention tools, as well as groups that train activists in using circumvention and security technologies, also competed for the earmarked funds.

The GIFC found powerful allies in Mark Palmer, former US ambassador to Hungary when the Iron Curtain fell, and Michael Horowitz, a former Reagan administration official and longtime advocate for human rights and religious freedom. They argued that if the GIFC could get sufficient funding to scale up its tools, authoritarian regimes would be brought to their knees—and that therefore failure to fund GIFC represented gross negligence as far as human rights were concerned. After the State Department, charged with disbursing the money, chose in 2008 not to award any of that year's funds to the GIFC, Horowitz and Palmer launched a lobbying and media campaign, urging editorial writers and reporters in major newspapers including the *Washington Post*, *New York Times*, and *Wall Street Journal* to promote the GIFC's technologies and to question the State Department's competence as well as that of the other organizations that had received funding. In their view, the failure to fund the GIFC was proof of the State Department's diplomatic squeamishness about supporting a religious sect that the Chinese government considers a direct enemy.

Though the State Department did grant a small amount of funding to the GIFC in 2010, in January 2011 it announced that a remaining $28 million in unspent funds would support not only circumvention tools, but also technologies and organizations aimed at helping Internet users in repressive countries defend against a much broader range of threats, including total Internet shutdown (most famously deployed in Egypt but also used by other governments including Syria, Burma, and selectively by China in specific regions and locations); aggressive denial-of-service attacks on activist websites; spyware installed surreptitiously on activists' and journalists' computers; hacking of activists' social media accounts; and the deletion by Internet companies of sensitive content, deactivation of accounts, and tracking of user behavior.

In retaliation against the State Department for failing to support circumvention tools exclusively, Senator Richard Lugar, ranking Republican on the Senate Foreign Relations Committee, called for the remaining funds to be removed from State Department control and given to the Broadcasting Board of Governors (BBG), the government agency that runs the Voice of America, Radio Free Asia, Radio Free Europe and Radio Liberty, Al Hurra TV, and other US-government funded broadcasters. The BBG had been funding circumvention tools including the GIFC to help their audiences access their websites, and unlike the State Department, they committed to spend the funds exclusively on circumvention. In April 2011 after aggressive lobbying by GIFC supporters, Congress decided to split the difference, cutting the State Department's annual funding for Internet freedom technologies and training by one-third and giving that portion to the BBG, which was expected to continue funding the GIFC, among others.

In the end, roughly half of the State Department funding went to circumvention technologies, with the rest of the money spent on other projects, including mobile security technology and training on protecting against hacking and online surveillance, as well as a project that is sometimes called "Internet-in-a-box" or "Internet-in-a-suitcase"—a set of tools and technologies for people to remain connected, at least in a rudimentary, ad hoc way, when networks get shut down.

Bizarrely, most of the people involved in the fight for Internet freedom funding said little and did nothing about a blatant contradiction: although US taxpayer money is being spent to help activists get around censorship, much of the censorship in North Africa and the Middle East is being carried out largely with North American software—as the OpenNet Initiative's report on the sale of North American censorship technologies to repressive regimes has documented. This technology is "dual use" in the sense that it can also be used by families to protect children from inappropriate content or dangerous contact with ill-intentioned adults, or by network administrators to defend against attacks, but the extent to which companies are marketing their tools to repressive regimes—with full knowledge of how they will be used—has been the subject of much less energy and discussion in Washington than the issue of who ought to be receiving Internet freedom funds. Moreover, the fight over circumvention funding only further distracted politicians, policy makers, media pundits, and journalists from the deeper question of what Internet freedom actually means.

GOALS AND METHODS

Writing in *Foreign Affairs* in late 2010, New York University's Clay Shirky critiqued Washington's obsession with circumvention. Such an "instrumental" approach, he argued, is counterproductive in the long run. The main problem that circumvention technology aims to address is the blocking of content and platforms based *outside* a country, run by people over whom the government in question has no direct jurisdiction or control. Circumvention is much less helpful to people seeking to create their own content and build their own locally based information communities and networks. "The potential of social media," Shirky wrote, "lies mainly in their support of civil society and the public sphere—change measured in years and decades rather than weeks or

months."This was certainly the case for the Egyptian and Tunisian revolutions: journalists and policy researchers deconstructing events after
the fact discovered that activists in these countries had spent years building not only an online community skilled in the use of social media, but
also a network of offline ties and trusted relationships. Similarly, the efforts by beleaguered liberal Chinese bloggers and intellectuals, who
struggled for nearly a decade to build and nurture both online and offline spaces for liberal-leaning, political criticism of the kind not permitted on Chinese commercial platforms, have been largely quashed by
the Chinese government.

Concretely, if one applies Shirky's framework to the Chinese situation, one can see how the government's attack on a community of
liberal-minded bloggers through arrest, threats, and harassment—a
community with both online and offline components that took nearly
a decade to build and deepen—is many magnitudes more insidious and
harmful than the blockage of websites run by the Voice of America
and Radio Free Asia, or even YouTube, Facebook, and Twitter. Instead
of viewing the Internet and social media as instrumental to freedom,
Shirky advocates what he calls an "environmental" approach, focused
on the idea that "positive changes in the life of a country, including
pro-democratic regime change, follow, rather than precede, the development of a strong public sphere."This requires a fundamental shift in
strategy, from providing uncensored access to outside content to "securing the freedom of personal and social communication among a
state's population, [which] should be the highest priority, closely followed by securing individual citizens' ability to speak in public. This
reordering would reflect the reality that it is a strong civil society—one
in which citizens have freedom of assembly—rather than access to
Google or YouTube, that does the most to force governments to serve
their citizens."

Others argue that the US government's Internet freedom policy
should not treat the Internet as an instrument for regime change—
especially given that some groups and governments elsewhere in the
world are using the Internet to help weaken, challenge, or destabilize

the American system of government. Furthermore, as Ethan Zuckerman of Harvard's Berkman Center warns, if US policy approaches the Internet as if its main value is as a conduit for the Voice of America, repressive regimes will be more likely to treat American Internet companies—and the Western technology industry more generally—as enemy combatants. US-based Internet companies' local employees as well as their most active local users will more likely be viewed as foreign agents. Rather, Zuckerman argues, the goal should be much broader and ideologically agnostic: "to ensure that people can make their voices heard in this new space, and hope that governments will be wise enough to listen and to engage."

Writer and critic Evgeny Morozov has been even more critical of the US government's "Internet freedom" policy, warning that any policy based on the assumption that the Internet inherently helps democracy and hurts authoritarianism is misguided, counterproductive, and downright dangerous. His book *The Net Delusion* offers a scathing condemnation of the "cyber-utopian" and "Internet-centric" worldviews he believes to be epidemic among American academics (including Shirky and Zuckerman), alongside many activists, foundations, journalists, politicians, and investors. He mocks what he calls the "Google Doctrine—the enthusiastic belief in the liberating power of technology accompanied by the irresistible urge to enlist Silicon Valley start-ups in the global fight for freedom." Cyber-utopianism, he argues, is dangerous because it fails to recognize that the Internet "penetrates and reshapes all walks of political life, not just the ones conducive to democratization." The Internet, he points out, empowers dictators, demagogues, and terrorists as much as it empowers democrats. How the Internet interacts with politics and the particulars of how it is used for good and for ill vary drastically from country to country.

US Internet freedom policy also has critics among its intended beneficiaries overseas. Tunisian blogger and activist Sami Ben Gharbia—who was heavily involved in the movement to bring down the Ben Ali regime—wrote a passionate essay in September 2010, just four months before his government fell, explaining how US government involvement in grassroots digital spaces can be counterproductive by endangering people who are already vulnerable to being accused by nasty regimes of being foreign agents and unhelpfully causing authoritarian governments to view Western Internet companies as their enemies. He quoted the Egyptian blogger and activist Alaa Abd El-Fattah, who argued that Western democracies must sort out their own domestic obstacles to Internet freedom if they want to be genuinely helpful to Middle Eastern Internet freedom and democracy in the long run:

> Fight the troubling trends emerging in your own backyards from threats to Net neutrality, disregard for user's privacy, draconian copyright and DRM [digital rights management] restrictions, to the troubling trends of censorship through courts in Europe, restrictions on anonymous access and rampant surveillance in the name of combating terrorism or protecting children or fighting hate speech or whatever. You see these trends give our own regimes great excuses for their own actions. You don't need special programs and projects to help free the Internet in the Middle East. Just keep it free, accessible and affordable on your side and we'll figure out how to use it, get around restrictions imposed by our governments and innovate and contribute to the network's growth.

Despite critiques from activists like Ben Gharbia and Abd El-Fattah and intellectuals like Morozov, and the challenges brought by WikiLeaks' release of confidential State Department cables at the end of 2010, Clinton was determined to press forward with her Internet freedom policy. In February 2011, at the height of the Arab Spring, she gave a second Internet freedom speech. Though events in the Middle East and North Africa seemed to have vindicated her department's belief in the Internet as a key policy priority, gone was the Churchillian tone and the Cold War metaphors of the previous year's speech. She admitted that neither she "nor the United States government has all the answers," or even all the right questions. Whether the

Internet—the "public space of the twenty-first century"—is used positively or negatively, she noted, depends on each and every one of the world's two billion–plus Internet users, as well as all governments who seek to regulate it, and companies that build Internet technologies and platforms.

Still, the State Department has been unable to escape the contradictions between US Internet freedom policies and Washington's pursuit of national security, counterterrorism, trade, and copyright interests. Many people around the world are cynical about these contradictions in precisely the same way that people are cynical about how US economic and security interests regularly contradict, and sometimes override, the degree to which the United States is willing to emphasize democracy and human rights in its diplomacy and assistance priorities with particular countries. Examples abound: In early 2011 protesters in several countries reported that, after being teargassed by riot police, they found empty gas canisters marked "made in USA" lying in the street. While the Bahraini government was arresting bloggers and suppressing dissent, the United States was planning to sell $70 million in arms to Bahrain. When Clinton visited Cairo a month after the revolution, Egypt's January 25 Revolution Youth Coalition refused to meet with her, because "the US administration took Egypt's revolution lightly and supported the old regime while Egyptian blood was being spilled."

Despite these hypocrisies and contradictions, the Obama administration is moving to reinforce, broaden, and institutionalize its Internet freedom policy, first championed by Clinton. In May 2011, the administration published a document called the "International Strategy for Cyberspace," outlining the US government's overarching goals in preserving the global Internet as an open, interoperable, secure, and stable network. "While offline challenges of crime and aggression have made their way into the digital world," wrote President Obama in the introduction, "we will confront them consistent with the principles we hold dear: free speech and association, privacy, and the free flow of information." The challenge now is for American citizens—and netizens everywhere—to hold the US government to this commitment. As Part

Three of this book discussed in detail, that will not be easy. Still, that the administration has made a free and open Internet into an official policy goal at least makes it difficult for US officials to ignore or discount public criticism of any US government actions that undermine Internet freedom.

DEMOCRATIC DISCORD

In the months following Clinton's first speech in January 2010, Internet freedom quickly became a buzzword in foreign ministries around the democratic world, especially in northern Europe. Swedish Foreign Minister Carl Bildt called for a "new transatlantic partnership for protecting and promoting the freedoms of cyberspace." His ministry then proceeded to take the lead in facilitating several international meetings over the ensuing year and a half, including two multi-stakeholder brainstorming conferences on global Internet freedom.

Positive momentum continued to build through the summer. In July 2010 the French and Dutch foreign ministers convened an international conference on the Internet and freedom of expression attended by representatives of seventeen governments, in addition to dozens of representatives from nongovernmental organizations, businesses, and international organizations. They announced agreement on several points, including that the international community must improve monitoring actions of governments to ensure that they protect—and avoid violating—online free expression; that there must be better mechanisms for holding companies accountable when they collaborate with repressive censorship and surveillance; and that more must be done to "come to the aid of cyber-dissidents." They also agreed to hold two more meetings later in the fall.

Unity crumbled before any further meetings could take place. Less than a month before the next conference, planned for Paris in late October, President Nicolas Sarkozy declared in a speech at the Vatican, "Regulating the Internet to correct the excesses and abuses that arise from the total absence of rules is a moral imperative." Soon thereafter, the French free speech organization La Quadrature du Net published a leaked letter from Sarkozy to French Foreign Minister Bernard Kouchner, in which Sarkozy described the conference as an "opportunity to promote the balanced regulatory initiatives carried on by France during these past three years, and in particular the HADOPI law in the field of copyright, which has recently been supported by the European Parliament, as well as the measures taken to fight the new phenomena of cybercriminal."

Sarkozy, it turned out, intended to use—or to put it more bluntly, hijack—the Paris conference to advocate a Europe-wide version of his "three strikes" law to fight online piracy, despite widespread controversy about the extent to which it erodes legal due process, among other concerns. Dutch Foreign Minister Uri Rosenthal responded with a statement that the Netherlands does not support such laws and that he no longer planned to attend. The conference was "postponed" and never rescheduled. Jérémie Zimmermann, spokesman for La Quadrature du Net, called his president's attempt to hijack the conference "one more example of the alliance between the entertainment industries and a few politicians, who seek to control the public space to remain in power."

Sarkozy was more successful at promoting his agenda eight months later, at a gathering called the "e-G8": an exclusive conference of Internet CEOs, government representatives, and assorted Internet celebrities, organized by the French public relations firm Publicis and held as a prelude to the annual G8 meeting scheduled that year for Deauville, France. Addressing the Internet executives and CEOs in attendance, including Google's Eric Schmidt, Amazon's Jeff Bezos, and Facebook's Mark Zuckerberg, Sarkozy declared, "The world you represent is not a parallel universe where legal and moral rules and, more generally, all the basic principles that govern society in democratic countries do not apply." In a speech not long before the conference, he had said something similar: "The Internet is the new frontier, a territory to conquer. But it cannot be a Wild West. It cannot be a lawless place."

He spoke as if there were no middle ground between total anarchy and the particular solutions he favors. The problem with Sarkozy and too many other policy makers is not that they object to arrogant, self-ish, and irresponsible behavior of companies. The problem is that leaders such as Sarkozy offer a false binary choice between their pre-ferred solutions on the one hand and an anarchic state of nature in cyberspace on the other without allowing for any alternatives. Such a philosophy of Internet regulation is, essentially, neo-Hobbesian: strong nation-state interference to save people from chaos and crime, with the corollary that any rational, responsible citizen should be willing to give up his or her digital freedoms to a trusted authority who knows best.

For the same reason that the world's political thinkers outgrew Hobbesian logic and moved on to Locke, "consent of the governed," and democracy, it is time to consign bipolar Sarkozian arguments for Internet regulation to the dustbin of history. To anybody who believes in democracy (as opposed to anarchy, dictatorship of the proletariat, theoc-racy, enlightened technocracy, bonapartist autocracy, or some other ap-proach to governance) Internet freedom does not mean Internet anarchy or vigilante justice any more than physical freedom in the democratic context means absence of government. The difference between Internet freedom and Internet tyranny is not *whether* the Internet should be gov-erned; instead it is a question of *how* the Internet should be governed. An Internet that is compatible with—and conducive to—democracy should be governed in publicly accountable ways that reflect the will and respect the rights of the governed. This approach will in turn re-quire an equitable balance of power between government, corporations, and citizens.

Fortunately, some European leaders are still seeking alternatives. In mid-2011 after holding a large consultative conference earlier that spring, the Council of Europe published two documents: a set of prin-ciples on Internet governance and a declaration on the Internet's "uni-versality, integrity and openness," beginning with the assertion that "the right to freedom of expression is essential for citizens' participation in democratic processes. This right applies to both online and offline ac-tivities and is regardless of frontiers." The objective of the two documents is to ensure that at least within Europe, government laws and regulations, international treaties, and corporate business practices must all be compatible with these fundamental principles to ensure that citizens' rights and interests are not ultimately harmed.

In June 2011, UN Special Rapporteur on Freedom of Expression Frank La Rue delivered a report to the UN Human Rights Council that not only condemned the censorship and surveillance practices of authoritarian countries, but also warned of dangerous trends in the democratic world that threaten citizen rights in the Internet age. He pulled no punches in his critique of efforts by many democratic governments to hold intermediary companies liable for the actions of their customers and users on the grounds that they are in effect delegating the role of censorship and surveillance to unaccountable private actors. "Holding intermediaries liable for the content disseminated or created by their users severely undermines the enjoyment of the right to freedom of opinion and expression," he wrote. "It leads to self-protective and overbroad private censorship, often without transparency and the due process of the law."

La Rue stressed the need to preserve citizens' right to online anonymity as a prerequisite for dissent and whistle-blowing, calling on governments to refrain from requiring "real name" registration on social networks, as in South Korea. He was also "deeply concerned" and "alarmed" by French and British "three strikes" laws. Cutting off Internet access as a response to copyright infringement, he wrote, is "disproportionate and thus a violation of article 19, paragraph 3, of the International Covenant on Civil and Political Rights." Neither La Rue nor the UN Human Rights Council (of which forty members endorsed La Rue's report) has the power to compel any government to do anything. However, the moral weight of the Human Rights Council has nonetheless helped to shine a global media spotlight on a broad range of governments that threaten Internet freedom. Citizens' groups in various parts of the world will use the La Rue report as a powerful tool in pushing for more reasonable human rights and democratic approaches to Internet regulation.

CIVIL SOCIETY PUSHES BACK

On the eve of the May 2011 e-G8 meeting in Paris, former Grateful Dead lyricist and Electronic Frontier Foundation cofounder John Perry Barlow sent out a tweet quoting Sarkozy: "the Internet is a new frontier, a territory to conquer." Then he added, "And I am in Paris to stop him."

"For the first time in human history, it is possible to convey to every human being the right to know . . . and the right to express him or herself," Barlow later declared from the stage. "This is a very important legacy to give to our children, and if we are going to deny them that legacy on the basis of trying to preserve some old institutions that have outlived their usefulness, we will not be good ancestors." The audience erupted with cheers and applause. Several other academic and activist speakers, invited primarily due to their status as Internet celebrities, pointed out that the people who will drive the Internet's future—users from the developing world, Middle Eastern cyber-dissidents, and the young programmers launching start-ups from their bedrooms—were conspicuously absent, destroying any pretense that the meeting represented the interests or values of anybody other than one group of elites.

Technology blogs and Twitter networks lit up with clever one-liners in defense of an open and free Internet. La Quadrature du Net organized an ad hoc press conference at the end of the meeting to reiterate opposition to Sarkozy's approach. In the end, the calls of activists, academics, and Internet entrepreneurs for an open Internet with minimal regulation dominated news headlines about the meeting. Sarkozy's effort to build consensus around his vision had not gone exactly as planned—because in this important instance, people took action to defend netizen rights and succeeded in bending history, at least a little bit. This success should be an example for other small efforts that can defend an open and free Internet at a time when a series of mostly disconnected initiatives and decisions are determining its fate.

Just as the Tunisian and Egyptian cyber-activists did not spring immaculately from Twitter and Facebook, the La Rue report on the threats to online free expression did not spring forth from a well-oiled

UN research apparatus. Though he drew upon statistics compiled by UN sources such as the International Telecommunication Union, government reports, and information supplied by Internet companies, he also relied heavily on research and policy papers compiled by civil society groups and academics. La Rue also held five consultative meetings in different parts of the world. He heard directly not only from Internet experts, academics, and corporate representatives but also from local activists and online journalists who provided him with first-hand descriptions of the obstacles they face when trying to use the Internet to organize, access information, and disseminate their own reports and ideas.

Also in early June 2011, UNESCO published a report titled *Freedom of Connection—Freedom of Expression: The Changing Legal and Regulatory Ecology Shaping the Internet*. Many of its revelations and concerns were similar to La Rue's, with an added emphasis on a reality that needs to be understood by citizens as well as by companies and governments. "Freedom of expression is not an inevitable outcome of technological innovation," the report concluded. "It can be diminished or reinforced by the design of technologies, policies and practices—sometimes far removed from freedom of expression." Because the Internet is globally interconnected and national governments have failed to offer regulatory solutions that respect and protect the rights of all netizens, they stressed the need for "a stronger multi-stakeholder framework for Internet governance at the international level," with a special emphasis on including free expression and human rights activist groups in the process of setting rules for a new global digital environment. The authors called for the creation of a "special international taskforce for freedom of expression" to better "support and represent these stakeholders in Internet governance."

At a conference soon after the report's release, the Council of Europe's commissioner for human rights, Thomas Hammarberg, accused UNESCO of dodging responsibility. He argued that the United Nations should handle such matters directly rather than handing them off to a task force. As the issue of Internet and human rights moves into the mainstream and comes into focus for many politicians for the first time, people like Hammarberg are apparently unaware of the history behind UNESCO's recommendation. Frank La Rue did indeed issue an excellent report and set of recommendations—thanks to substantive engagement and hard work in collaboration with academics and activist groups from around the world. But counting on the United Nations as an institution to take practical, concrete steps to protect human rights online has already been proven to be counterproductive.

CHAPTER 13

Global Internet Governance

By 2005, Tunisia under President Ben Ali had already gained a reputation as the Arab world's leading Internet censor. That the United Nations chose Tunis as the venue for a global conference on the future of the Internet could not have been more ironic—or symbolic—of why putting the UN in charge of coordinating the Internet's practical functions would be a setback for freedom of expression.

The World Summit for the Information Society, or WSIS, had three components: first an intergovernmental negotiation over who controls the organizations that coordinate the Internet's global functions; second, a nongovernmental forum with panels and workshops organized by citizens' groups around the theme of "ICT4D," or how to use Internet and communications technologies (ICTs) for economic development; and third, a commercial trade fair in which technology companies from around the world could promote themselves to a global clientele. China's two largest networking companies, Huawei and ZTE (China's second-largest telecommunications equipment maker), were the main corporate sponsors.

Tunisian security forces predictably cleared the roads near the conference hall of ordinary Tunisians. Opposition groups were barred from the proceedings and a number of local activists were arrested. Attempts to hold a parallel international citizens' forum in another part of the city were quashed. The head of the French press freedom group, Reporters Without Borders, was barred from the country. Chinese leaders prevailed on Ben Ali to ban the New York–based Human Rights in China from entering Tunisia. Of course, the Internet in the conference hall—as in the rest of the country—was heavily censored. Human rights groups pointed out that the UN's choice of Tunisia as host was yet another example of its tendency to lend legitimacy to dictators. It also reinforced the view of many people in the democratic world as well as the global human rights community who believe that control over the Internet's core functions and technical standards should be kept out of the UN's hands.

Thus we face a challenge: if civil liberties are to be preserved and protected, Internet governance cannot be left up to governments alone to negotiate and coordinate among one another. Exactly what the alternative should be, or how it should work, or what success looks like all remain open questions.

THE UNITED NATIONS PROBLEM

In November 2005, I was invited to WSIS by a Dutch foundation to help moderate a discussion and workshop at the nongovernmental forum. Titled "Expression Under Repression," our panel included bloggers from Iran, Zimbabwe, China, and Malaysia; a report on Tunisian Internet censorship by a technologist from the OpenNet Initiative; and a hands-on tutorial, organized by my colleague Ethan Zuckerman, for activists wanting to learn how to use the Internet anonymously and evade censorship. Though the UN organizers had approved the workshop and issued conference credentials to all our invited participants, the Tunisian government tried to cancel our event. It did not appear on the official schedule and Tunisian officials informed our workshop's organizers that it was "off topic." Free expression, the Tunisian hosts said, had nothing to do with "ICT for development," the official theme. In the end, the Dutch ambassador had to intervene so we could proceed.

At the appointed time, we found our designated room packed with Tunisian plainclothes police and members of official Tunisian news organizations. As soon as the panelists finished speaking and we moved into a question period with the audience, a woman from Tunisian state television got up and gave a long speech in French about the arrogance of rich Westerners who lecture people in poor countries about human rights. Countries need to resolve basic issues of food, shelter, and connectivity before they can consider "luxuries" like human rights. I asked our panelists to respond. Activist Taurai Maduna from Zimbabwe silenced the naysayers with a pointed comment: "If we have no freedom of speech, we can't talk about who is stealing our food."

On the other hand, events in Tunisia also underscored just how uneasy many non-Western, developing-world governments feel—including many democratic ones—about a new digital infrastructure that they and their people increasingly depend on, but that is largely engineered and coordinated by people in the economically prosperous West. It is a fact of life that people who have been living in the economically prosperous democratic West all of their lives (no matter how well-meaning they may be) have difficulty understanding and anticipating other people's linguistic and cultural identities—let alone economic and political aspirations. This tension leads to suspicion about what the West's real motives are when it comes to Internet governance and whether the real motives are more about preserving the West's dominance in the global marketplace of goods and ideas than about freedom and democracy that conflicts with those aspirations. This tension is real and needs to be taken seriously if the Internet is going to develop in a manner that will serve the fastest-growing segment of Internet users—who happen to live in non-Western, largely non-English-speaking countries.

This East-West imbalance was one of the many reasons cited by the Chinese government in its 2005 bid to remove control over critical Internet infrastructure from US government and private, largely Western, hands. Up until the late 1990s, the system coordinating Internet domain names and IP addresses—a system that ensures, for example, that when a person types www.cnn.com into the address bar on a browser, that person gets the same website everywhere—was handled through a largely informal process by a group of engineers living mainly in the United States. In 1998, this coordination function was formalized with the establishment of a San Diego–based nonprofit corporation called the Internet Corporation for Assigned Names and Numbers, ICANN. Its budget comes primarily from the fees collected through the sale of domain names. Technical and policy decisions are made by a board of directors in consultation with various "constituencies" of network engineers, corporate interests (companies that either administer, sell, or use domain names), Internet user groups, and governmental representatives, through an elaborate system of councils meant to balance out the interests of the Internet's different stakeholders.

Under ICANN's original governance structure, the US government had no role in the organization's day-to-day decision-making. But because the Internet had by then become so important for the American economy, ICANN was required to sign a memorandum of understanding in which it committed to ultimate oversight by the Department of Commerce. This relationship led to mounting concerns by other governments over the past decade as the Internet became more important—indeed critical—to more countries. Many governments were uncomfortable with the idea that the United States had direct power over such a vital if still young global system. Given the frequent tensions plaguing US-China relations and the strategic emphasis it places on technology, the Chinese government was particularly uncomfortable. But so too were the Russians and the Iranians, as well as others including democracies such as Brazil and even some in Europe.

In the run-up to the 2005 Tunis meeting, the Chinese government led a bid to dismantle ICANN and transfer its functions to a UN body, the International Telecommunication Union (ITU). Many developing-world governments supported this plan because they have more power in the ITU, where all countries have equal voice, no matter whether they are represented by engineers or by diplomats who may or may not understand the technical complexities under discussion. At ICANN, the perspectives of Western engineers and Western companies have tended to dominate, especially in its early years—and those lacking technical understanding have been paid little attention.

Despite recognizing that this digital divide does not serve the interests of developing-country Internet users, human rights and civil liberties groups from all over the world have been united in opposition to any plan to transfer control over critical Internet resources from ICANN to the United Nations. Such a move would empower governments that routinely practice political and religious censorship. Dictatorships would gain greater influence over the Internet's regulatory and technical norms in ways that could potentially undermine or imperil dissenters and unpopular minorities who use the Internet all over the world. The United States stood with the human rights groups, which was not difficult since human rights principles conveniently overlapped with the US self-interest in keeping ICANN under its umbrella rather than handing it off to the UN. The Europeans sought a middle ground by which ICANN's oversight would be more international but not directly under UN control either.

In the end, diplomats at the World Summit agreed to disagree and to maintain the status quo. ICANN was left alone, but with an admonition to become more international and inclusive, which it has subsequently attempted to become by bringing more governments into its multi-stakeholder consultative policy process. Meanwhile, the UN delegates agreed to create a new talk-shop, the Internet Governance Forum (IGF), where governments, companies, and NGOs from all over the world now meet annually to discuss Internet policy issues.

Though the IGF has no power to set policy or make binding decisions, it does provide a forum through which governments, companies, and NGOs from all over the world can identify common problems, disagreements, and solutions. Over the past six annual IGF meetings, many transnational "public-private partnerships" have emerged to handle issues like cyber-crime, the digital divide, and child protection. In his book *Networks and States*, Internet governance scholar Milton Mueller of Syracuse University argues that the IGF, if organized and managed well, has the potential to serve as a global "coral reef," supporting a whole ecosystem of formal and informal "policy networks." Through these networks that are formed or broadened through IGF participation, professionals working for different governments, companies, and NGOs can forge ties that enable them to solve problems for the world's Internet users in flexible, ad hoc ways.

One of the IGF's many problems is that it is dominated by whichever governments, companies, and organizations have the financial resources to participate. Governments from the developed world are well represented at the IGF, while participation by those from poorer countries is generally patchy. Governments with clear agendas on Internet governance issues such as China, Iran, India, and Brazil position themselves as representing the interests of the non-Western "Global South." Corporate interests are mainly represented by the major Western multinationals plus a few other large players from China, Japan, and India. Nongovernmental interests are represented primarily by international NGOs based in Western countries, plus some from India and a few of the better-resourced democratic African countries with active technical and engineering communities, such as South Africa, Ghana, and Kenya. Nongovernmental voices from authoritarian countries are almost entirely absent. As a result, only the governments—or government-approved so-called nongovernmental organizations—speak for the interests of Internet users in those countries.

In 2009 the IGF was held—ironically once again—in Sharm El-Sheikh, Egypt, the resort city where former president Hosni Mubarak maintained a home and where he and his wife were held under house arrest since the February 2011 revolution. Little did the IGF participants imagine that in just over a year, Mubarak's regime would be brought to its knees by Internet activists—none of whom were invited to the meeting.

An entire morning of the four-day forum was disrupted by an unscheduled speech given by First Lady Suzanne Mubarak, who wanted to promote her Cyber Peace Initiative for online child safety. Intel was a major sponsor of her initiative. Family safety organizations from the United States and Britain rushed to sign partnership agreements with Mrs. Mubarak's organization. At one point during the meeting I asked members of one prominent British child safety organization if they knew about the Egyptian government's record of arresting and torturing young Egyptian bloggers and Facebook users critical of Mubarak. They were not. I asked if they knew that child safety is commonly used by authoritarian regimes as an excuse for censorship and surveillance.

Two people who play leading roles in their organization expressed great surprise and discomfort. It was clear that they had not considered these issues, or how they might inadvertently be lending legitimacy to a regime that does not respect human rights.

As a speaker on a panel about social networking services, I planned to cite Egypt's record of jailing and torturing bloggers, China's system of corporate-level censorship, and South Korea's real-ID requirements, among other cases. Just beforehand, Nitin Desai, a UN undersecretary-general and chairman of the Internet Governance Forum, took me aside and warned me not to mention any UN member countries in my remarks. Otherwise, since this was a UN meeting he would have to give all governments the right of reply, which would use up all available time and prevent the panel discussion from proceeding. Earlier during the conference, the OpenNet Initiative had been confronted by UN security after it tried to display a banner in the hallway advertising a launch event for *Access Controlled*, a book about global censorship. According to the security guard who removed their poster, a certain UN member country had complained because the ONI had not obtained permission from the IGF secretariat to display the poster.

As an organization formed to coordinate policy between nation-states, the United Nations has clearly struggled to run a process aimed at ensuring that the Internet evolves in a manner that benefits all stakeholders—an Internet that people around the world can reasonably trust to protect their rights or defend their interests. The most problematic aspect of the IGF is the charade perpetuated by many of its most powerful participants that it actually is a neutral and fair discussion platform through which governments, companies, and citizens' groups can discuss common problems of the digital realm. So far it has fallen well short of this description.

ICANN—CAN YOU?

In December 2010, ICANN CEO Rod Beckstrom gave an impassioned speech at the United Nations defending the multi-stakeholder model as more compatible with the Internet's intrinsic nature than any kind of global governance system based on nation-states. ICANN, Beckstrom argued, is proof of how well "the multi-stakeholder model works." He was certainly right that the nation-state system is not the appropriate framework for governing the Internet. The truth is, however, that ICANN's multi-stakeholder process is an early-stage experiment. ICANN works much better for some countries, companies, and Internet users than it does for others. Though its multi-stakeholder decision-making model means that even some of the world's most powerful governments do not always get their way, ICANN has not yet figured out how to serve the interests of all the world's netizens fairly and efficiently. Until it does so, its claim to sovereignty over even one specific function—coordinating the Internet's critical addressing and numbering resources—will remain vulnerable to attack from many different sides.

In June 2009, I attended one of ICANN's public meetings, held three times a year for a week in a different part of the world. This one was in Sydney, Australia. Every meeting includes a public comment session: an opportunity for anyone attending the meeting (which is open to literally *anybody* who signs up) to raise concerns and questions directly to the board of directors. Twenty board members sat on a raised platform at one end of the Sydney Hilton's grand ballroom, gazing out from the dais at several hundred men and women, many sitting with laptops open.

A dozen or so people lined up behind two microphones placed in the aisles, taking turns to ask questions and make statements. Eventually an American woman with close-cropped silver hair took her turn. "My name is Marilyn Cade," she said. "I'm speaking as an individual and have appeared before this council of master Yodas many times. . . . I use the term master Yodas because I think, in fact, I understand that we are asking you to exercise wisdom."

Cade worked for AT&T for many years and is now an independent consultant who is heavily involved with the politics of both the IGF and ICANN. The specific issue she raised that day had to do with the board's right to appoint a panel of experts to study a question related to intellectual property and domain names. But the larger point is that, apart from the absence of any known extraterrestrials, her analogy between the ICANN board and the *Star Wars* Jedi Council was apt. The men and women on the dais are stewards and protectors, not of "the Force," but of the Internet. Unlike the Jedi Council, they are appointed with term limits. Like the Jedi Council, the source of their power and legitimacy is fuzzy. Their authority will last only as long as the bulk of the world's network engineers and most governments choose to respect it.

ICANN's function is narrow but critical. It runs the world's domain name system, or DNS. Most people who use the Internet do not need to know anything about the DNS, thanks to the people and organizations around the world who make it work smoothly. Web pages, e-mail, and other applications hardly just float in space; their data actually resides on computer servers physically located somewhere. When you type a web address into your browser, you expect to get the same website no matter what kind of browser or device you are using, and no matter where in the world you happen to be. When you send an e-mail to a particular address, you want it to go to the same person no matter where in the world you're sending it from and where that person is located. Every computer server or network has an IP address. The trick to making everything work is to make sure that all of the domain names used by everybody all over the world all correspond to the same IP addresses. If there is a discrepancy, then a single globally interoperable Internet will be "broken."

IP addresses also have to be allocated. The original system, called IPv4, provided roughly 4.3 billion unique numbers (less than the world's population). Nobody imagined when the Internet was created that those numbers would run out, but they did in 2011. A global transition effort is now underway to replace the old system with a new IP addressing system, IPv6, with more than enough numbers for every molecule on earth. Somebody has to allocate and keep track of the distribution of IP addresses—whether they are "v4" or "v6." Currently that job is done by an organization tied to ICANN called the Internet Assigned Numbers Authority (IANA), which allocates blocks of IP addresses to five regional Internet registries (RIRs), one for each of five regions of the world. The

RIRs are run by networking engineers from all the countries in each corresponding region; large organizations such as universities, banks, and Internet service providers (ISPs) and any other organization requiring large numbers of IP addresses pay to become members of their local RIR. These RIRs hold regular meetings, at which engineers discuss technical problems and make consensus-based decisions about solutions. The global system works because all of these engineers have agreed to trust one another and honor their regional RIR's decisions.

Though governments and government-affiliated organizations around the world are involved with the RIRs and engineers from all over the world help run them, the system is not controlled by governments—with one big exception: IANA is under contract with the US Department of Commerce. Herein lies the core issue that many other countries, especially China, object to so strenuously. There are periodic moves to wrest control over IP address allocation and domain name system coordination away from IANA, the other RIRs, and ICANN and put these functions under governmental control through the UN's International Telecommunication Union or similar. The UN system's chronic inability to protect and uphold human rights around the world, and its propensity to empower and legitimize dictators within the global governance system—as well as the lack of technical understanding of how the Internet really works among many countries' ministers of communications—are good reasons that power over the Internet's critical resources should be kept out of intergovernmental hands.

But as the Internet's importance grows, and as a critical mass of the world's Internet users expands far beyond its original core of affluent Westerners to include much less affluent people on mobile devices, and people who read and write only Chinese, Arabic, Urdu, Japanese, Korean, Russian, one of India's eighteen official languages, or anything else that doesn't easily "interoperate" with the English-language alphabet, the strains and stresses of Internet governance are mounting. ICANN is struggling to manage these strains in a way that facilitates the Internet's development, and in a way that genuinely works well for all of its users, not just for the wealthiest English-speaking ones.

Why should people who are not network engineers care about these seemingly obscure issues of Internet governance? Most of the world's Internet users have never heard of ICANN, the DNS, or RIRs, or any of the global power struggles taking place within and around these and many other acronyms. The outcome of these power struggles, however, will affect the extent to which dissent and unpopular speech—or any speech that displeases powerful governments or large brand-name corporations—can have safe passage and a safe home on the Internet. Many of the actors in these power struggles claim to be representing the interests of people who have no idea that they exist, and who likely would not trust them if they were aware.

A lot of the recent politics around the DNS have to do with who controls domain names and who makes money from them. Until 2010, all domain names were in Roman letters only—if your native language was Chinese or Arabic, tough luck, you still needed to be sufficiently familiar with the Roman alphabet and to know how to type Roman letters or you couldn't use the Internet. That started to change in 2011 when ICANN rolled out what are called "internationalized domain names," or IDNs. Every country gets what's known as a country-code top-level domain (ccTLD in ICANN acronym-speak), such as .cn for China or .uk for the United Kingdom or .ca for Canada. Starting in 2011, countries with the technical ability to manage the process could also register IDN ccTLDs. Egypt and several other Arabic-speaking countries rolled out Arabic top-level domains so that businesses and organizations in their own countries could register web addresses entirely in Arabic, making them accessible to more people for whom the English alphabet is completely alien. China, Korea, and Japan have all rolled out similar IDN variants of their country-code top-level domains. Since domain names are sold to individuals and corporations for money, this linguistic expansion of course is a great business opportunity for a whole industry that has grown up since the 1990s around enabling people to purchase and trade domain names.

Domain names like sex.com and business.com have sold for millions of dollars, benefiting the tech-savvy first-movers who grabbed lucrative domains in the 1990s before most businesses realized how important these names would become for them. In June 2011, after several years of disputes and bargaining among the organizations and various constituencies, the ICANN board decided to expand the Internet's real estate even further by allowing any organization with enough money and technical capacity to apply for new generic top-level domains (gTLDs). Companies will be able to create top-level domains for their brand (.ibm, .apple) and cities can create top-level domains for themselves (.seattle, .berlin, .beijing)—and all of these things can be done in any major language or script on earth. ICANN will charge a five-figure registration fee for the privilege of running and administering a top-level domain name, plus annual fees on top of that.

This is a lucrative opportunity for all concerned: it gives people an opportunity to create virtual real estate out of nothing and then sell it. For example, a business or organization serving people surnamed MacKinnon could in theory (assuming they can afford the $185,000 application fee and a further annual fee of $25,000) apply to run the .mackinnon gTLD. That way, instead of buying rebeccamackinnon.com, I could purchase http://rebecca.mackinnon from a domain name registrar. If a coalition of people of Scottish descent wanted to apply for ownership of the .mac gTLD as a community-building or moneymaking venture, however, we would run into trouble. Apple owns the trademark for "Mac."

Control over property and real estate in the physical world has always been the focus of complex and heated politics—in fact, resistance to the king's arbitrary power to confiscate land and property was the main impetus of the Magna Carta and subsequent basis of "consent of the governed" in modern times. Power struggles over control of land and property have always been at the core of politics and geopolitics. Now those struggles have entered a new dimension with the Internet.

Digital real estate is turning out to be as political as physical real estate, with the added complication that the network is borderless and global, which means that the politics of digital domains are waged simultaneously within countries and across many borders. The political and commercial clashes of the physical world are remixed and recast in complicated ways. Some kinds of businesses and other constituencies of Internet users stand to gain from the expansion of Internet domain names; others believe they have much to lose as ICANN moves ahead with the new gTLD program.

Chief among the potential losers are large Western multinational companies with internationally famous trademarks and brands, which lobbied heavily against the domain name expansion. Back in the 1990s, there were famous cases of "cyber-squatting," whereby tech-savvy and opportunistic individuals registered domain names of famous brands, followed by .com, and demanded high sums of money from the companies. In other cases, legitimate disputes arose over who had the right to a particular domain: Volkswagen went to court over vw.com, which had been registered first by a small company in Virginia called Virtual Works. The court ruled in Virtual Works' favor. Thus many name-brand companies fear that a potentially unlimited number of new gTLDs will bring new costs to "own" all domain names related to their brands, compelling them to act quickly every time a new gTLD is launched, buying up thousands more domain names related to their brands, and when necessary suing for ownership of domain names they believe are rightfully theirs.

After the ICANN board decided in 2008 to begin laying the technical, legal, and political groundwork to launch new gTLDs, an intense battle ensued over who has a right to apply for any given name in any language, how the application process should be run, what criteria should be used for acceptance or rejection, and who, if anyone, gets veto power. Governments concerned about the use of certain names to promote dissident platforms or content deemed offensive demanded veto power over applications. Companies concerned about protecting their trademarks and brand names (known within the ICANN world as the "intellectual property" constituency) advocated strict controls preventing anybody except brand owners from registering any names in any language that resemble their brands. Many Western governments, lobbied heavily by those companies, supported their position.

On the other side of the argument, civil liberties groups and consumer rights advocates countered that giving veto power to governments and corporations over new domain names would chill free expression. Companies and governments from the developing world argued that excessive trademark and copyright protection measures being pushed by major Western multinationals would discriminate against less globally famous, non-Western companies with a following in their own communities but whose brands may have names similar to those of major multinationals.

ICANN's board makes the final decisions, but it takes policy recommendations from a number of councils and supporting organizations that deliberate via conference call, e-mail, and in-person meetings. These councils and supporting organizations consist of people claiming to represent different constituencies (noncommercial users, consumer groups, commercial registrars and registries, trademark-holding companies, engineering interests, etc.). Governments participate in an advisory role only, through what is called the Governmental Advisory Committee (GAC). No government representatives sit on ICANN's board of directors, which has demonstrated clear independence from the world's most powerful governments, including the United States. In June 2011 the United States and the European Commission led an effort to force ICANN to delay the launch of the new gTLD program, due to what they felt were unresolved trademark and commercial concerns. The ICANN board, confident that the rules set out for the new gTLD program were the result of a consensus developed over years by a diverse set of stakeholders, ignored the objections from Washington and Brussels and approved the program anyway. Among those applauding ICANN's defiance was Eliot Noss, CEO of the Canadian company Tucows, which sells domain-name, e-mail, software, and small business services, who commented after the board vote that "if you're talking about the Internet, nations and nation states are just actors at the table, not predominant."

Non-Western governments also complain that ICANN's multistakeholder structure contains a Western-centric bias because the organization is dominated by people from the Western developed world. The working language is English, with representation by businesses and nongovernmental organizations too geographically narrow to reflect the full range of concerns and challenges of Internet users in developing countries where most of the Internet's expansion is now taking place. Because of this imbalance, a growing chorus of governments—with China in the lead—have demanded that ICANN elevate the role and influence of the GAC in the organization's decision-making process so that their people's interests can be better represented.

Unfortunately, the authoritarian and quasi-authoritarian governments active in the GAC are even less interested in defending the right to free expression and privacy of their citizens than the Western governments are. At least businesses and nongovernmental groups from democratic countries can openly oppose their own governments' positions at ICANN. Entrepreneurs and civil society groups from countries where open defiance of the government is not tolerated cannot come to ICANN and advocate for positions that clash with their government's. Thus, even though ICANN claims to be a multi-stakeholder body with a bottom-up approach to its decision-making and policy-making processes, in reality the interests of Internet users from authoritarian countries are represented only by governments, businesses with close governmental ties, and quasi-governmental organizations. The interests and rights of dissidents, politically unrepresented minorities, and cyber-activists from nondemocratic countries have no meaningful representation at ICANN from any quarter, except indirectly from the very few international human rights and free-speech groups with the staff, resources, and expertise to engage in ICANN policy-making processes.

In 2009 Global Voices, the grassroots citizen media network that I cofounded, became a member of the Non-Commercial User Constituency of ICANN's Non-Commercial Stakeholder Group (NCSG). The NCSG was created to represent the interests and defend the rights of people around the world who use the Internet largely for noncommercial purposes, which of course includes political activism. Ironically, noncommercial users have had to fight hard within the ICANN bureaucracy and arcane governance structures to gain fair representation within what is billed by its leadership as a bottom-up, grassroots, and inclusive decision-making process.

In 2009 the NCSG was told that in exchange for six seats on one of the stakeholder councils that make policy recommendations to the board, it would be allowed to elect only three of those seats, with the remaining three noncommercial user "representatives" appointed by the board. The NCSG was not allowed to write its own governance charter but instead had one imposed by ICANN staff. "Welcome to 'bottom-up' policy making at ICANN," remarked Robin Gross, constituency chairperson at the time. Milton Mueller reported on his blog, "Apologetic Board members openly confessed that they did this simply to appease the commercial user groups who, they feared, would 'go ballistic' and create trouble for them in Washington if they did not."

Over the ensuing two years, representatives of the beleaguered non-commercial constituency—most of whom participate in ICANN as un-paid volunteers, unlike corporate representatives to ICANN whose companies pay for their participation and consider it part of their jobs— negotiated and bargained with the board and ICANN staff for a more favorable charter, including the right of its members to elect all of their representatives. The NCSG also continued to challenge the ICANN board and staff regarding weak representation from "developing and transitional" countries. In response to such challenges from NCSG and others, ICANN has improved its "remote participation" tools, which enable people from around the world to follow and participate in meetings through the Internet and by phone. ICANN has also expanded a fellowship program for "individual members of the Internet community who have not previously been able to participate in ICANN processes and constituent organizations." Much more will need to be done, however, if the interests of all the world's netizens are to be fairly represented at ICANN.

Certainly, when it comes to upholding human rights and free expression, the UN governance model is far worse. It is clear that it is not in the interest of the world's netizens to leave Internet governance to nation-states. Yet the structures and processes that have so far been built for multi-stakeholder Internet governance are failing to mediate the kind of global politics needed to uphold and protect human rights, civil liberties, and free expression in the global network.

Meanwhile, ICANN lumbers on, despite accusations from all sides of ineffectiveness, waste, mismanagement, and questionable legitimacy. Milton Mueller points out that multi-stakeholder institutions like ICANN have their own form of international, multi-stakeholder "pluralist politics" that allow for various stakeholders to speak and be heard and to lobby for their interests—assuming they can afford to participate both in terms of time and in terms of resources to attend meetings (and the political risk if they are trying to represent nongovernmental interests of people in authoritarian countries). The problem is that this multi-stakeholder political process takes place without a basic values framework—at a national level often called a constitutional framework—that would prevent political outcomes from violating the rights of some Internet users in various parts of the world, because either they are on the losing side of a bargaining process or their concerns are not even represented or understood. Mueller rightly concludes, "There can be no cyberliberty without a political movement to define, defend, and institutionalize individual rights and freedoms on a transnational scale."

The next step is to build that movement.

CHAPTER 14

Building a
Netizen-Centric Internet

Almost six months to the day after the ouster of Tunisian president Zine El Abidine Ben Ali, a small protest of roughly 150 people formed in the middle of Tunis.

"We are all Samir Feriani," the protesters chanted, brandishing photos of the forty-four-year-old police officer. Feriani had been arrested two weeks previously after writing to Tunisia's interior minister, naming several high-ranking ministry officials who he said were responsible for killing protesters and committing other human rights abuses during the Tunisian revolution. In one of the two letters published in a local magazine, he further claimed that ministry officials had been destroying sensitive archives since Ben Ali's ouster, including archives of the Palestinian Liberation Organization (based in Tunis from 1982 to 1994), which he said documented Ben Ali's relationship with Israeli intelligence. Feriani was charged with "harming the external security of the state," "releasing and distributing information likely to harm public order," and "accusing, without proof, a public agent of violating the law."

Tunisia's twitterati clamored to support Feriani, dismayed that elements and behaviors of the old regime lingered on in the new Tunisia. "Where are our journalists, the civil society and the political parties?" asked a Twitter user called @emnamejri. On Facebook, people created pages calling for his release. They posted pictures of protests, links to news about his case, and aggregated reactions of citizens around Tunisia. They circulated the condemnation by Human Rights Watch, which summed up many people's feelings about his arrest: "At a time when many Tunisians believe that the officials who terrorized people under Ben Ali remain strong within the security establishment, the provisional government should be encouraging whistle-blowers, not using the ousted government's discredited laws to imprison them."

At a conference in New York about the Internet and politics that same month, somebody asked Riadh Guerfali—the activist who made the 1984 Ben Ali mash-up back in 2004—whether he was worried that the Tunisian revolution would not end well. "I am optimistic that we will win this battle," he said. "At the present time there is still lots of trouble," he continued. "But public opinion is here." Despite the steep uphill battle, the difference between then and now is that Tunisians have carved out a space in the media and on the Internet for discourse and debate that is vastly broader and more accessible than under Ben Ali. "If we can say, this is wrong, we can say the person responsible must re-sign," that is the first step, Guerfali told the room full of young Ameri-can political operators, bloggers, journalists, and techies. "Things never ever, anywhere in the world, change by itself. It takes the pressure of public opinion."

This truth applies to everyone, everywhere. Democracy will not be delivered, renewed, or upgraded automatically, like the latest Netflix blockbusters through our broadband connections and smart phones. The future of freedom in the Internet age depends on whether people can be bothered to take responsibility for the future and act. Just as our individual actions and choices as citizens, parents, teachers, employees, managers, and government officials combine to shape the kind of world we live in, the actions and choices of each and every one of us are shap-ing the Internet's future.

Elements of a transnational movement to defend and expand Internet freedom have begun to emerge. Like the Internet itself, this movement is decentralized, loosely if at all coordinated, and driven often by groups and individuals at the edges reacting to specific problems. For now the movement is neither sufficiently broad nor sufficiently powerful to keep the abuse of power by governments or corporations systematically in check. But the revolutions, and attempted revolutions, of early 2011 have jolted many more people around the world into becoming actively engaged with the power struggle for freedom and control of the Internet.

What should this movement be aiming for? Establishing some sort of global UN-like uber-government to manage and restrain cross-border digital power is neither realistic nor desirable. Nor is Robin Hood–style cyber-vigilantism and digital guerrilla warfare. Given human nature and the realities of today's world, it is also inconceivable to expect to start completely afresh with some sort of utopian digital democracy in a pristine and politically unspoiled frontier of cyberspace.

A more realistic and democratic approach is to build and strengthen alternative netizen-driven institutions and communities that can exist alongside existing ones, eventually shifting the balance of power both online and off. At the same time, we must devise more effective and innovative ways to constrain all forms of digital power within reasonable limits, whether that power is exercised by governments, corporations, or activist hacker networks of varying ideological and religious stripes. The first step is to build much broader public awareness and participation. People need to stop thinking of themselves as passive "users" and "customers," and start acting like citizens of the Internet—as "netizens."

STRENGTHENING THE NETIZEN COMMONS

Rosental Alves, a Brazilian journalist who now heads the Knight Center for Journalism in the Americas in Austin, Texas, likes to compare the pre-Internet age to a desert. Most people had easy access to a very limited number of sources for news and information. People's understanding of the world depended on the priorities and budget decisions of the editors who ran news organizations, and whatever or whoever could influence them.

Then came the deluge. Today we live in an informational rain forest that sprang up around us practically overnight. We were not prepared for such an overwhelming ecosystem of rapidly evolving info-organisms. We have moved abruptly from a problem of scarcity to a problem of overabundance, at least for some varieties of information. Others are scarce and harder to find, drowned out or buried amid the rapidly proliferating dominant species. The bulk of online media are now things that we produce ourselves—on social media, on blogs, on personal websites. Journalists and media professionals now compete for attention and influence one another. "The logic of communication that is being created in this new era is based on engagement," Alves says.

Many elected officials of the world's major democracies complain about the low quality, viciousness, or navel-gazing nature of much of the discourse they see online. What many of the complainers seem not to recognize is that unless they want to destroy freedom of speech and anonymity on the Internet through overregulation and unrealistic demands on companies to police the citizen-created content they host and transmit (in which case they are enemies of freedom regardless of whether they intended to be), the only answer to bad speech is better and more effective speech.

Of course, just because anybody can now create and transmit media does not automatically mean that human society will be more democratic or peaceful. Life in the rain forest is just as likely to be nasty, brutish, and short, a Hobbesian state of nature that is not only primeval but also primitive. In the offline world, this is why we build civilizations. It is now up to the world's netizens to figure out how to build a sustainable civilization within the new digital rain forest—in which we find sustenance and shelter amid the poisonous plants and deadly predators.

Success is by no means inevitable. People still need to learn how to participate constructively and responsibly in this new space and protect the rights of minorities and dissenters. The right incentives and disincentives for good and bad behavior online have yet to be worked out. Solutions that adequately protect netizen rights will come about, however, only if netizens of the world participate actively in devising them. The more we actively use the Internet to exercise our rights as citizens and to improve our societies, the harder it will be for governments and corporations to chip away at our freedoms, arguing as they so often do that we do not deserve them, and treating us like reprobates.

Global Voices is one of many emergent communities of people who assert their citizenship of the Internet, not as passive users or consumers. They take personal responsibility for the future of online information—and its freedom—by contributing directly. There are many other such communities dedicated to building a new kind of civilization that will be better suited to our young and rapidly evolving digital rain forest, communities that do not insist upon maintaining the old desert ways of life, along with all the institutions and customs that made sense in the old context but may doom us in the middle of a deluge. Wikipedia, the encyclopedia that anybody can edit, is perhaps the most famous example of an information resource that anyone on earth can contribute to, but that is maintained and governed by a core community of people around the world who believe in the idea that elites no longer control human knowledge. They have proven that a group of people with a common set of objectives can govern themselves according to a set of community rules and produce a resource that no amount of public funding or corporate revenue can come close to creating.

Equally important are the more locally focused citizen media organizations dedicated specifically to document, call attention to, and take action against violations of citizens' digital rights. The Tunisian website Nawaat.org, run by Sami Ben Gharbia, Riadh Guerfali, Slim Amamou, and others, served as an anchor for the Tunisian digital activist community. Though they also used commercial platforms like Facebook, Twitter, YouTube, and Flickr as key components of their activism strategy, Nawaat.org served as a one-stop consolidator and archive of information. It also served as an important backup when commercial services were blocked or accounts were hacked or shut down. In Egypt, websites like the Torture in Egypt blog and the Egyptian Blogs Aggregator run by open-source programmer Alaa Abd El-Fattah similarly served as anchor points for more diffuse campaigns over multiple social media sites, which had broader reach but over which the activists themselves had less control in terms of how the information was stored and shared.

Another type of organism in the new ecosystem is Creative Commons, a nonprofit founded by Lawrence Lessig—then at Stanford, now at Harvard—dedicated to helping people share information and media, as broadly or as narrowly as they would like. Its flexible system of copyright licenses enables organizations like Global Voices, Wikipedia, and many other nonprofit citizen projects to ensure that their content is shared as widely as possible and translated into as many languages as possible, with the creators' full approval and consent. The point is that people who want to protect their works with traditional copyright certainly can, but many other people who create media for reasons other than sales should also be able to do so.

Other groups are focused explicitly on keeping the Internet as open and free as possible. In 2009, Mozilla (creator of the Firefox browser and other open-source tools) launched Drumbeat, a platform through which people can become actively involved in keeping the web open and free by organizing their own projects and recruiting others to help. One team is working to develop a set of universal "privacy icons"—a set of symbols that companies could adapt, which would make it easier for users to understand what personal information is being stored and for how long, and how and with whom it might be shared, under what circumstances. That in turn would make it easier for a user to decide how to use—and not use—any given service. Several other projects focus on creating free courses, instructional manuals, and other educational tools so that nontechie web users can improve their web literacy and ability to build and modify one's own tools. The point of the Drumbeat movement is that the web will be free and open only if people participate actively in making it so.

Still others focus on educating netizens about how they can protect themselves from threats to their freedom. Global Voices Advocacy works with a range of other nonprofit organizations to disseminate information in a range of languages about the threats citizens face to their freedoms and rights online, and what tools and tactics they can use to protect themselves. The Tactical Technology Collective develops training materials for citizen privacy and security, while Mobile Active in turn works with a network of technologists and activists to help people fight censorship and surveillance on mobile phones.

EXPANDING THE TECHNICAL COMMONS

[handwritten margin note: AREA WHERE PEOPLE GATHER]

When the Egyptian government shut down the Internet on January 27, 2011, a worldwide community of activist programmers and engineers— "hacktivists"—sprang into action. Internet and mobile service providers in Egypt were down, but as long as there were phone and fax machines capable of making and receiving international calls, there were still ways for Egyptians to connect to the Internet. The most famous effort was a collaboration between Google and Twitter, called "Speak to Tweet": people could find a landline, call a phone number, and record a message that would then be disseminated to the world through Twitter.

[handwritten margin note: WANTS CITIZENS TO KNOW WHAT IS OUT THERE (OF TOOLS)]

Other efforts actually enabled Egyptian activists to connect directly to the Internet despite the blackout. Members of a loosely organized group called Telecomix, originally formed in Sweden and now with hundreds of active members around the world, along with its operational sister organization called We Rebuild—both dedicated to promoting "access to a free Internet without intrusive surveillance"—began collecting information about dial-up Internet services in Europe and other countries in North Africa and the Middle East that Egyptian people could access from any landline with international service. It was an expensive call, but at the height of a political crisis it was better than nothing. The hacktivists then searched the Internet for Egyptian fax numbers and began faxing the information as far and wide as they could. They also put up a website on which they updated new dial-up information regularly, and distributed that to expatriate Egyptians, who then called the landlines of friends and relatives, who could then pass on the information. A small French ISP offered the free use of its dial-up service by Egyptians. Other such accounts were purchased from services around Europe and North America by supporters of the Egyptian revolution living around the world.

A range of activist entrepreneurs, nonprofits, and foundations are now looking for ways that activists can be more prepared, whenever and wherever the next government flips the "kill switch." Soon after the Egyptian shutdown, a consortium of developers, led by the New America Foundation's Open Technology Initiative, began integrating a number of existing open-source platforms into easy-to-use software that can be installed on almost any device with a Wi-Fi connection, so that in the future when a government blacks out the Internet, activists will at least be able to remain connected to one another by linking together their laptops and Wi-Fi-enabled mobile phones in a local "mesh" network. Dubbed Commotion Wireless, a core component of the project is a software program called Serval, which enables people to create an ad hoc independent cell phone network using their existing phone numbers, even when the normal commercial networks are destroyed or switched off. A Serval-based network has already been deployed successfully as a test case in the Australian Outback. Data are stored within the local network for relay onto the Internet when and where a connection can be established at least occasionally. (The project received funding from the State Department in spring 2011.)

The idea of mesh networking is not new. Civic-minded engineers and software developers have been working for more than a decade to address the problems caused by the combination of government and corporate control over both how, and whether, ordinary citizens can access the Internet. The community wireless movement first emerged in North America and Western Europe as part of an effort to provide access to remote and economically disadvantaged citizens whom corporate Internet and wireless carriers have little or no business incentive to serve. The largest and most successful community wireless project is Guifi.net, which started in Catalonia and has expanded into other parts of Spain's Iberian Peninsula where rural households have not been reached by Telefonica, the main Spanish telecommunications company. Other projects are under way in Berlin, Vienna, and Athens. In Detroit an organization called the Digital Justice Coalition is promoting community-owned broadband and wireless services that prioritize locally produced content as a way to promote community, local innovation, and civic involvement. An important component is local mesh networking within neighborhoods: one resident or business obtains a commercial Internet connection, then that signal is shared or relayed throughout the neighborhood by inexpensive routers in each house. A whole community of software and hardware innovators has emerged around these projects, sharing software code and hardware designs, and learning from one another's successes and failures.

In authoritarian states and bonapartist quasi-democracies, community-controlled connectivity is difficult to establish, let alone maintain without reprisal, and is thus likely to emerge only in situations involving open revolt and rebellion. In democracies, local connectivity movements are still so small and new that their broader political implications remain untested. Proponents of community wireless and neighborhood mesh networks are working to ensure that laws, regulations, and technical standards can enable community wireless to thrive and coexist alongside—and even sometimes in symbiotic relationship with—existing corporate services. In a time when it is becoming increasingly difficult for people to participate fully in politics and political discourse, start a successful small business, or avail themselves of government services without Internet and wireless connectivity, we netizens should be free to organize alternative means of connectivity when commercial options fail to meet our needs.

For people seeking to evade censorship and surveillance on whatever network they happen to be using, activist software developers have been working for years on a range of tools. A range of software developers with different commercial and non-commercial affiliations have developed software tools that help Internet users in China, Iran, and other authoritarian countries to access blocked content. Some are more effective and secure than others. One group of open-source engineers has spent the past decade working on an "anonymizer" tool called Tor, which enables users to surf the web and upload or download content without being traced. Between the start of the Egyptian protests on January 25, 2011, and the blackout on January 27, the use of Tor from Egypt more than quadrupled, due to activists' concern about police surveillance of their Internet communications. Developers are also working to bring secure Internet access with Tor and other tools to mobile phones, or at least those using Google's open-source Android operating system. Some countries, such as Iran and China, have figured out how to block Tor's publicly known channels, creating an arms race in cyberspace between activist developers and government engineers.

Other activist engineers and programmers are working on the problem of corporate control of our social and political lives online. In 2010 at the height of negative publicity about Facebook's privacy policies, four young programmers from New York University's Courant Institute launched an open-source project called Diaspora, which aims to provide a software download or "pod" that people can install on their own computer servers to create their own Facebook-like social networks. Rather than having Facebook controlling people's personal information and policing users, the idea is for decentralized groups to be able to control their own platforms and data. Their online drive to raise $10,000 that summer garnered so much interest that they ended up raising nearly $200,000. A group of programmers in the Seattle area has developed open-source social networking software called Crabgrass, tailored specifically for the needs of political activists. An older project, StatusNet, enables people to set up their own Twitter-like microblogging services that they can control locally.

In early 2011 Columbia University professor Eben Moglen announced a new project dubbed FreedomBox, aimed at addressing the vulnerability of activists and dissidents who currently rely too much on centralized corporate services like Facebook and Amazon to store their data and disseminate their media messages. The problem, he explained in a speech, was caused by the creation of a standard service architecture that is "very subject to misuse" because it consists of "vast repositories of hierarchically organized data about people at the edges of the network that they do not control." When the police show up with a warrant at the offices of the company that controls the network, the company must hand over the data. Moglen's idea is to create a network of cheap and small servers, no larger than a cell phone plug that would be locally controlled by individuals, linked together as a decentralized network, so that citizens' data cannot be acquired by authorities or any other powerful entity from any one centralized place. Such a system would enable people to back up their data, publish, and communicate securely—all anonymously if they so wish.

As of mid-2011 Diaspora and FreedomBox remained in developmental and experimental stages. StatusNet was being used by a number of tech-savvy communities but has so far failed to gain widespread traction as a noncommercial alternative to Twitter. It remains unclear whether large numbers of people will ever be interested in switching from large commercial brand-name services to more secure and locally controlled alternatives.

Wide adoption of noncommercial tools and services is certainly not unheard of, however, if they are user-friendly enough and offer clear value for many people. Nonprofit organizations, individual activists, independent media organizations, and low-budget educational institutions all over the world rely on WordPress, the open-source blogging platform, and Drupal, the open-source content management system, both of which are developed, maintained, and upgraded by a community of volunteer developers. Another extremely successful project dedicated to developing open-source software that helps ordinary nontechie Internet users gain greater control over their online lives is Mozilla. Its Firefox browser, an open-source volunteer-developed web browser that allows for a high degree of customization, now makes up roughly 30 percent of the world's web browser usage. Firefox integrates the work of other developers through add-ons and plugins, including many that help increase people's privacy and security online. One of these is the Torbutton, integrated with Tor, which enables users to surf the web anonymously and circumvent blocked websites. The Mozilla community also works on a range of other software tools, including the open-source e-mail client Thunderbird, which has become the preferred e-mail client for many activists because it is easy to encrypt, in addition to being free.

Even though a handful of the projects listed above (particularly Commotion Wireless and many but not all of the circumvention tools) do receive some government and corporate support, many of the program-mers and engineers involved in building the global technical commons are part of a hacker counterculture that distrusts all authority and all in-stitutions. This culture is epitomized by groups like the Chaos Com-puter Club (CCC), which became famous soon after it was founded in Berlin in 1981 after some of its members hacked into the German post office's computer network to expose the network's inadequate security measures. Once every four years the group holds an event called the Chaos Communication Camp in an open field. Generators are set up to power an ad hoc wireless network; workshops are held in tents. The 2011 schedule included topics like "Running Your Own Services to Im-prove Privacy"; "How to Set Up Your Own Not-for-Profit ISP"; "Blackout Resilient Technologies"; and "How to Bypass the New Data Retention Law." These extraordinary summer hack-fests are comple-mented by yearly winter conferences the CCC website describes as a "diverse audience of thousands of hackers, scientists, artists, and utopi-ans from all around the world." It is at such events that many of the new tools and techniques of digital resistance are first tested and deployed.

UTOPIANISM VERSUS REALITY

In 1996, John Perry Barlow famously wrote a manifesto titled "A Dec-laration of the Independence of Cyberspace." It began, "Governments of the Industrial World, you weary giants of flesh and steel, I come from Cyberspace, the new home of Mind. On behalf of the future, I ask you of the past to leave us alone. You are not welcome among us. You have no sovereignty where we gather." In the sixteen years since, government has certainly not left "us" alone in cyberspace—not in small part because many of "us" sought government help in defending us from the criminals, pedophiles, bullies, industrial spies, racists, ter-rorists, and others who have extended their activities into cyberspace. Meanwhile, corporations have built their own private sovereignties that sometimes challenge—and sometimes collude with—government sovereignties.

Some activist programmers and intellectuals now believe the solution to the twin ills of government interference and unaccountable corporate conduct is to abandon the government- and corporate-addled Internet and build a new one—or a new extension of it—that can be independent of both. In a widely quoted January 2011 essay, media theorist Douglas Rushkoff called on the netizens of the world to unite: "Instead of pretending that the Internet was ever destined to be our social and intellectual commons," he wrote, "we can much more easily conspire together to build a real networked commons, intentionally. And with this priority embedded into its very architecture and functioning." Realizing such an ideal, however, is problematic on both practical and ideological levels.

On a practical level, activists cannot wait for the ideal world to be built. The point of activism is to reach, convince, and engage the largest number of people as quickly and effectively as possible. As Clay Shirky has pointed out, technology becomes most powerful only after it has become commonplace. "The invention of a tool doesn't create change," he writes in *Here Comes Everybody*. "It has to have been around long enough that most of society is using it." Ethan Zuckerman's "cute-cat theory of digital activism" makes a similar point, based on his own experience over years running both commercial and nonprofit platforms. He observes that online activism is most effective when it is carried out through platforms and services that were created not for earnest and civic purposes, but rather for frivolity and fun: online spaces that most people use to socialize, follow and discuss sports teams and movie stars, show off pictures of their babies, and—naturally—trade silly photos of each other's cats. The most popular destinations for online fun and frivolity are of course commercially operated.

Creating noncommercial digital spaces tailor-made for activism has other challenges, as the group WITNESS discovered firsthand. Launched in 1992, WITNESS is a Brooklyn-based nonprofit organization whose primary mission is to help organizations use video to document human rights abuses and advance human rights causes. For the first decade of the organization's existence, it focused mainly on training activists in video shooting and editing, and in serving as a bridge between activist groups and the global media. By 2005, with the launch of YouTube and other video-sharing sites, it was clear that commercial video-sharing platforms were a powerful tool for activists. But there was a problem: commercial platforms failed to address human rights activists' need for safety, security, and privacy.

As an answer to this problem, in 2007 WITNESS launched its own Video Hub—a sort of YouTube for human rights activists, if you will, tailored to activists' concern for privacy and security, among other needs. They ran the Video Hub for two years before deciding to close it down. One reason was that the technical challenge and expense of hosting vast amounts of video, as well as keeping the site user-friendly and defending it from attack, became insurmountable for a nonprofit organization. The second reason had to do with audience. "Very practically," wrote WITNESS Director Yvette Alberdingk Thijm, "this means that we will more proactively go where people are, as opposed to asking them to come to us."

WITNESS shifted to a new model of curating video posted by activists on other video-sharing websites, most of them commercially operated. That did not, however, stop them from trying to address the problem that inspired them to start the Video Hub in the first place: commercial platforms continue to fail regularly in addressing activists' needs, risks, and concerns. WITNESS has worked directly with YouTube to help its staff develop better, more activist-friendly policies and procedures to minimize the chances that activists' videos would not get removed due to misunderstandings by company "abuse" teams about the nature of their video as well as misunderstandings on the activists' part about the company's terms of service. In 2010 the organization collaborated with YouTube on a guide to "Protecting Yourself, Your Subjects, and Your Human Rights Videos on YouTube." YouTube sent a senior executive to the 2010 Global Voices Citizen Media Summit in Santiago, Chile—a gathering of bloggers from all over the world—to explain its system and learn more about digital activists' concerns.

At various times, members of the Global Voices community have discussed whether it would make sense to develop a blog-hosting or so-

cial media platform for online activists, given all of the problems activists have faced with commercial platforms. In the end, however,
Global Voices decided that this was no more likely to succeed than
WITNESS had been, for the same reasons. Global Voices uses a combination of open-source tools, noncommercial platforms, and commercially operated services. This further reinforces how important it is that
the citizen commons—given its symbiotic relationship with the commercial sector—must engage and push companies to behave in a manner that is not only profitable for the long term but also serves the
greater public good.

Cyber-separatism has ideological dangers as well. Attempts have
been made over the past century to build government- and corporate-
free communities in the physical world. Most have turned out disap-
pointingly for most of the participants, who eventually returned to more
conventional lifestyles, failed to become economically sustainable, developed their own governance problems that led to bitter conflict, or all
of the above. The hippie communes of the 1960s were less comfortable
and harmonious in reality than in theory. Earlier in the twentieth century, revolutionary attempts to create capitalism-free societies in the former Soviet Union, Eastern Europe, China, and elsewhere were rather
disastrous when it came to human rights, let alone economic prosperity.
Utopian ideologies such as Marxism-Leninism and Maoism produced
demagoguery, totalitarianism, and genocide.

In a controversial 2006 essay about what he calls "Digital Maoism,"
and later in his 2010 book, *You Are Not a Gadget*, technologist Jaron
Lanier warned of a "new online collectivism," the digital variant of a
concept that "has had dreadful consequences when thrust upon us from
the extreme Right or the extreme Left in various historical periods."
Though there is much idealism and enthusiasm around the idea of the
Internet being a place where the evils, hypocrisies, and general messiness
of human economics, politics, and social relations can somehow be tran-
scended, there is little evidence that human nature is any more virtuous
or selfless in cyberspace than it is in the physical world. Yet there is copious evidence that the Internet can amplify and telescope both the

For example google in st [they have] a little power.

good and the evil aspects of human nature. Movements to create an ideal society through the creation of online communities led by charismatic leaders with utopian visions claiming to transcend all of the political ambiguities and hypocrisies of "meat space" are more than likely to produce Internet-age versions of the same problems that caused tremendous human suffering in the twentieth century.

A related danger is technological determinism: the belief that technology can be used to solve problems whose roots ultimately lie in human social and ethical behavior. Placing excessive expectations on the ability of technology to defeat repression can cause individuals to abdicate individual responsibility. Variants of technological determinism are often spouted by Internet company executives, who claim that by making the world more connected, they are inevitably and inexorably helping to make it more democratic in the long run, whatever the short-term compromises or collateral damage might be. This argument is wearing thin against attempts such as the Global Network Initiative to convince companies that they need to allow themselves to be held publicly accountable if they want the public to trust them over the long term.

Technological determinism is as dangerous as historical determinism, the worldview underpinning the philosophies of Marx and Hegel, who believed that history was inevitably and inexorably moving the human race toward a certain endpoint. Marxist revolutionaries believed they were in the vanguard of the historically inevitable. Karl Popper, in Volume Two of *The Open Society and Its Enemies*, his seminal 1945 defense of liberal democracy, warned that historicism "is in conflict with any religion that teaches the importance of conscience." Real human progress, he argued, can be achieved only "by defending and strengthening those democratic institutions upon which freedom, and with it progress, depends. And we shall do it much better as we become more fully aware of the fact that progress rests with us, with our watchfulness, with our efforts, with the clarity of our conception of our ends, and with the realism of their choice."

Certainly we can and should use technology to do precisely that. Just because revolutionary cyber-Soviets or Robin Hood–style cyber-vigilantism are not the answers to our problems does not mean that business and government as we know them today are serving the needs of today's citizens and netizens. New approaches to governance, accountability, and politics clearly are needed if democracy is to survive and thrive in the Internet age.

GETTING POLITICAL

It is not realistic for most people living in developed Western countries to live independently from corporate products and services. It is similarly unrealistic for all but the most dedicated and technically adept people to live their digital lives independent of corporate services. This is why political activism to push for netizen-centric corporate practices and government policies is essential. Companies did not adopt responsible environmental and labor practices of their own accord: they grew more responsible and accountable with their environmental and labor practices as a result of many decades of activism, investigative journalism, public pressure, and debate. If it had not been for decades of such activism, governments would not have moved forward in these areas either.

Similarly, ensuring that the Internet serves netizens' aspirations for democracy and accountable governance—and is not used ultimately to quash and manipulate these aspirations—is going to require a robust and diverse ecosystem of efforts and organizations over many years. The Internet freedom movement has not even arrived at the same point of global public awareness that the environmental movement achieved by the first Earth Day in 1970. There is much work to do.

Within the global environmental movement, some organizations and initiatives have seen value in working with corporations and governments. Others are opposed to compromise and insist on radical alternatives as the only course. All points on the spectrum need to exist for progress to be made. Meanwhile, in universities, students are starting to create activist organizations. The experience with student-driven movements for South African divestment in the 1970s and '80s, Sudan divestment activism in the mid-2000s, and on-campus chapters of

Amnesty International, Greenpeace, the Sierra Club, and others shows just what a critical and catalytic role students can play in promoting change when they get organized. One new organization, devoted to promoting and building the digital commons, is Students for Free Culture. Perhaps it is time for some talented and passionate young people to take the lead in launching a new global alliance: Students for Internet Freedom, anyone?

Just as green parties have emerged over the past several decades from the environmental movement, and labor parties emerged from the labor rights movement of an earlier generation, a new generation has begun to organize political parties and focus political platforms with a strong focus on Internet rights. Branches of the provocatively named Pirate Party now exist in at least twenty-five countries and have gotten candidates elected to local office in Spain, Switzerland, Germany, and the Czech Republic. In 2009 Sweden's Pirate Party won two seats in the European Parliament with 7 percent of the Swedish vote thanks mainly to the support of people under thirty years old. When Amazon's web-hosting service dropped WikiLeaks as a customer, the Swedish Pirate Party welcomed WikiLeaks to its secure and surveillance-free ISP, launched in mid-2011 in defiance of Sweden's data-retention laws. The Pirate Party has its origins in battles over copyright law and enforcement, and in defending citizens' right to use file-sharing websites. However, the party's recent electoral success at least in Sweden is due to a broadening of its platform to the fight against censorship and surveillance, which appeals to a much wider range of voters. In other European countries, green parties have taken up Internet freedom as a signature issue alongside environmental and social justice concerns.

Still, much of the public discourse about censorship and surveillance is found mainly in online niche media, blogs, and social media networks, read by niche communities of tech-savvy young people. Unlike environmental and labor issues, about which there is broad global awareness, there is much less so of the threats to Internet freedom and how these threats cut across regions, ideologies, systems of government, and corporate platforms. Major news organizations generally treat Internet-related news as a business, consumer, or cultural subject; with a few notable exceptions, the Internet generally is not covered as a politically contested space in which citizens need to engage to ensure their rights are upheld. Rather than wait for this to change, people and organizations with expertise and experience to share can report directly through blogs and independent websites, and in this way influence public debates as well as the broader news agenda.

To this end, universities and research institutes are starting to build research programs across disciplines—from computer science and political science to business, economics, and sociology—that help policy makers, companies, and citizens better understand the threats to Internet freedom around the world and how they can be counteracted. Teams at Harvard's Berkman Center, the University of Toronto's Munk Centre, the Oxford Internet Institute, Princeton's Center for Information and Technology Policy, and many others have contributed much of what the world knows—and what the media reports—about global Internet censorship and surveillance, cyber-attacks against digital activists, corporate practices that both extend and diminish freedom, and how different laws can either protect or erode civil rights in the digital environment.

A growing number of nongovernmental organizations are also dedicating themselves to Internet policy advocacy: informing the public about complicated issues that the news media often doesn't cover well, and lobbying governments to change or improve laws so the Internet can remain as open and free as possible. The Electronic Frontier Foundation and the Center for Democracy and Technology are just two of many such groups in the United States. Counterparts exist all over the world: the Open Rights Group in the UK, Bits of Freedom in the Netherlands, Netzpolitik in Germany, La Quadrature du Net in France, Jinbonet in South Korea, and many others. Other more globally focused organizations such as the South Africa–based Association for Progressive Communications are working to coordinate policy strategy on a global level, lobbying the United Nations Human Rights Council, the

Internet Governance Forum, ICANN, the OECD, and other regional and international organizations.

Ad hoc coalitions of these groups are also working to involve more netizens from around the world in global Internet governance debates and processes. In an effort to build consensus around the common values that netizens' groups are fighting for, the Dynamic Coalition on Internet Rights and Principles—a multi-stakeholder group composed mainly of activists and academics from around the world who have been meeting annually at the Internet Governance Forum and working throughout the year mostly through e-mail and Skype—spent two years collectively drafting a Charter of Human Rights and Principles for the Internet. Their aim is not to invent new rights, but to take rights that have long been considered universal in the physical world and translate them into the digital context. In early 2011 in conjunction with the Internet freedom advocacy group Access Now, they published a summary of the charter's ten core principles:

1) *Universality and Equality*

   All humans are born free and equal in dignity and rights, which must be respected, protected, and fulfilled in the online environment.

2) *Rights and Social Justice*

   The Internet is a space for the promotion, protection, and fulfilment of human rights and the advancement of social justice. Everyone has the duty to respect the human rights of all others in the online environment.

3) *Accessibility*

   Everyone has an equal right to access and use a secure and open Internet.

4) *Expression and Association*

   Everyone has the right to seek, receive, and impart information freely on the Internet without censorship or other interference. Everyone also has the right to associate freely through and on the Internet, for social, political, cultural, or other purposes.

5) *Privacy and Data Protection*

Everyone has the right to privacy online. This includes freedom from surveillance, the right to use encryption, and the right to online anonymity. Everyone also has the right to data protection, including control over personal data collection, retention, processing, disposal, and disclosure.

6) *Life, Liberty, and Security*

The rights to life, liberty, and security must be respected, protected, and fulfilled online. These rights must not be infringed upon, or used to infringe other rights, in the online environment.

7) *Diversity*

Cultural and linguistic diversity on the Internet must be promoted, and technical and policy innovation should be encouraged to facilitate plurality of expression.

8) *Network Equality*

Everyone shall have universal and open access to the Internet's content, free from discriminatory prioritisation, filtering, or traffic control on commercial, political, or other grounds.

9) *Standards and Regulation*

The Internet's architecture, communication systems, and document and data formats shall be based on open standards that ensure complete interoperability, inclusion, and equal opportunity for all.

10) *Governance*

Human rights and social justice must form the legal and normative foundations upon which the Internet operates and is governed. This shall happen in a transparent and multilateral manner, based on principles of openness, inclusive participation, and accountability.

Many of these principles go far beyond what Western democratic governments are prepared to support. Some clash dramatically with the positions of many of the world's most powerful Internet and telecommunications companies. Even so, the process of formulating these principles demonstrates that when the Internet's future is viewed through a human rights and social justice lens instead of a commercial or national security lens, major disagreements can emerge even when governments and companies agree in principle on the need for an open, globally interconnected Internet as well as protection and respect for human rights and free speech.

In June 2011 the Paris-based Organisation for Economic Co-operation and Development (OECD) published a "Communiqué on Principles for Internet Policymaking," signed by forty governments and two nongovernmental stakeholder groups: one representing business and industry and the other representing the Internet technical community. The document was heralded by the United States and other governments as an important foreign policy consensus, at least among the world's major democracies, on core principles for safeguarding the open Internet and free flow of information, as well as the need for balance between the need to protect civil liberties and governments' need to provide security and protect business. Yet a third group that also participated in negotiations over the text of the communiqué—the civil society constituency whose members included groups advocating for free expression, privacy, and consumer rights—could not endorse the document.

Though the civil society members welcomed the communiqué's commitment to human rights, rule of law, and freedom of expression, they were unable to support it for two main reasons. The document, they said in their dissenting statement, placed excessive emphasis on cyber-security and intellectual property enforcement in a manner that could potentially be used to justify policy trade-offs that human rights groups believe to be unacceptable. An even greater concern was vague language calling for Internet intermediaries such as ISPs and social networking platforms to take on more responsibility in policing their services. They warned that Internet intermediaries "are neither competent nor appropriate parties" to make decisions about the legality of content posted or transmitted by their users. "Requiring them to make determinations on the legality of content or behaviour of users raises issues for transparency, due process and accountability and detrimentally impacts on citizens' freedom of expression."

It is unlikely that the OECD Internet principles would have been heralded as a step in the right direction if either the business or the technical community had refused to sign on. Despite the lack of support from groups concerned with human rights, free expression, and social justice, members of the Obama administration involved with Internet policy praised the principles as a way forward for governments to "address policy challenges" without violating fundamental rights. The document laudably helps to build a consensus among democracies around a shared commitment to open, globally interoperable Internet standards and the free flow of information—in stark contrast to the Internet governance policies of China and other authoritarian nations. At the same time, it shows that democratic governments remain in denial about an insidious reality at the digital intersection of governmental and corporate power: power over citizens' digital lives is being exercised in increasingly opaque and unaccountable ways.

Elected democratic governments are likely to remain in denial unless and until their electorates give them meaningful political incentives—both positive and negative—to change their priorities. Over time, the environmental, labor, and traditional human rights movements built powerful political constituencies that have shifted the domestic and international policy priorities of most of the world's democracies. There is no reason an Internet freedom movement cannot eventually do the same thing, with enough effort by enough people.

CORPORATE TRANSPARENCY AND NETIZEN ENGAGEMENT

The power of corporations to shape netizens' digital discourse and hence our political lives will not be constrained without new mechanisms and strategies for collective bargaining by netizens with corporations. The existing political and legislative processes of nation-states are failing to do the job. While it makes no sense for a company to try to duplicate the mechanisms of representative parliamentary democracy within their global constituencies, the status quo is also unacceptable. Netizens, companies, and governments all face an urgent moral imperative to innovate politically—in the broadest sense of the word—on a scale that matches the dramatic technical innovations of the past several decades.

Aside from the Global Network Initiative's nascent attempt to create a system of accountability for Internet and telecommunications companies, a few even more embryonic efforts to push companies in a more netizen-centric direction have begun to take shape. They will all require much broader global participation and awareness to be effective:

Boosting corporate transparency. This is essential to prevent the abuse of citizens' rights—be it by governments or by companies. As citizens, we have a right to know how our information is shared, with whom, and under what circumstances. We also have a right to know how our access to information, as well as our ability to disseminate it, is being shaped by any given service or platform. At the moment, this right is respected by few companies or governments.

Companies should be required to report regularly and systematically to the public on how content is policed, and under what circumstances it gets removed or blocked and at whose behest. In the summer of 2010, motivated by its commitments as a member of the Global Network Initiative, Google took a step in this direction by launching a website called the Transparency Report, tracking the numbers of requests it receives from governments to take down content or hand over user information, broken down by country. (Ironically, China had to be excluded because of Chinese state secret laws that could endanger local Google employees if the data were released.) With the caveat that requests from China prior to Google's withdrawal of its search engine are a black hole, the greatest volume of known requests in July through December 2010 came from democratically elected governments, with the US government well in the lead, followed by Brazil, India, and the United Kingdom.

Though the data are not complete and there are many unanswered questions, Senior Counsel Nicole Wong explained in a speech soon after the report's 2010 launch that Google's intent was to use data "as a basis to start a conversation about censorship and surveillance." She pointed out that "we see requests in almost every country in almost every election period." Another section of the Transparency Report tracks the accessibility of all of Google's services country by country, in close to real time. This tool enables people around the world to see which countries are deliberately blocking YouTube, Gmail, Blogspot, or any other Google services, as well as when traffic is cut off altogether. (Statistics quickly dropped to zero during the Egyptian blackout, for instance.) Governments should support this effort if they consider themselves in the least bit democratic.

Google has also invested in transparency tracking that goes beyond its own services. It has partnered with the New America Foundation's Open Technology Initiative on a project called the Measurement Lab, or M-Lab, an open platform that seeks to build tools and collect data that will help Internet users compare and evaluate the quality of broadband connections around the world. The tools include a "glasnost test," which can be used to check whether a given broadband connection is blocking or throttling performance of certain applications. A "mobile traffic test" enables users to determine whether a mobile Internet provider is discriminating between applications or services. Of course, it would be ideal if all mobile and broadband service providers around the world were more transparent with users about such information from the beginning. The hope is that projects like the M-Lab will push other companies in the same direction.

Netizens should demand that Internet and telecommunications companies follow Google's lead and improve upon it. All companies must publicly and clearly show how they gather and retain our information; how they share that information both with government and other companies; and in what way they may be shaping or prioritizing certain types of data over others. In doing so they can credibly demonstrate to us, as their constituents, that they recognize and take seriously the power they hold over us in the quasi-public spaces they operate, and they understand their duty to wield that power responsibly.

Building a more citizen-centric and citizen-driven information environment. As marketing guru and consumer advocate Doc Searls likes to say: "We have a submissive relationship with services." Internet and telecommunications services and platforms dictate the terms. We have no role in the conception or formulation of the terms; we are merely offered the choice of clicking "agree" or not.

"Individuals need to be at the centers of their own digital lives, and not peripheral dependents either of vendors or identity providers," Searls explains in the tenth-anniversary edition of a seminal book that he coauthored, *The Cluetrain Manifesto*, about how the Internet has turned markets into conversations. Searls and a group of software developers, businesspeople, and consumer advocates are working on a set of projects described as Vendor Relationship Management, or VRM. The aim is to empower netizens in shaping the terms of our relationships with companies. Specifically, the goals include:

1. Provide tools for individuals to manage relationships with organizations.
2. Make individuals the collection centers for their own data, so that transaction histories, health records, membership details, service contracts, and other forms of personal data aren't scattered throughout a forest of silos.
3. Give individuals the ability to share data selectively, without disclosing more personal information than the individual allows.
4. Give individuals the ability to control how their data is used by others, and for how long. This will include agreements requiring others to delete the individual's data when the relationship ends.
5. Give individuals the ability to assert their own terms of service, reducing or eliminating the need for organization-written terms of service that nobody reads and everybody has to "accept" anyway.

For such a world to be realized, companies and governments must actively support its development. As netizens we should demand that they do so.

Creating a more netizen-centric and netizen-driven information environment even holds lucrative opportunities for forward-thinking entrepreneurs and businesses. In January 2011 a report by the World Economic Forum declared personal data to be a new "asset class" and a "post-industrial" business opportunity for "a host of new services and applications" that can increase "the control that individuals have over the manner in which their personal data is collected, managed and shared." This insight is a classic example of what the *Harvard Business Review*'s Kramer and Porter called "shared value": the identification by forward-thinking entrepreneurs of business opportunities whose very purpose is to empower people.

Building processes for corporate engagement with users, customers, and other stakeholders. Anger over identity policy on Facebook and Google Plus, problems caused by Facebook's privacy policy changes, the disastrous Google Buzz rollout, and Flickr's clashes with activists might all have been prevented if companies adopted more innovative ways to engage with netizens around the world who are affected by their businesses. A way must be found for companies to work more directly with netizens, their constituents, in shaping products and services. Netizens need to devise more systematic and effective strategies for organizing, lobbying, and collective bargaining with the companies whose services we depend upon—to minimize the chances that terms of service, design choices, technical decisions, or market entry strategies could put people at risk or result in infringement of their rights.

The idea that companies can and should engage with user and customer communities is not new. In his book *Democratizing Innovation*, MIT professor Eric von Hippel documents an important trend in innovation: many of the world's most dynamic and innovative companies now involve customers and users directly in developing new products or improving on existing ones. Von Hippel found this to be the case in a range of industries, including medical devices, computer hardware and software, and a range of consumer products. It is now time for Internet and telecommunications companies to innovate politically, in the broadest sense of the word: build processes for engagement with users and customers, who are re-envisioned as constituents. Invent tools for bottom-up, constituent-driven innovation when it comes not only to developing profitable features, but also to anticipating threats to people's rights and civil liberties. Involve us directly in figuring out how best to mitigate those threats.

PERSONAL RESPONSIBILITY

Ai Weiwei, the outspoken artist who helped design the Beijing Olympic stadium before becoming a full-time thorn in the government's side, disappeared from the Beijing airport on April 3, 2011, as he prepared to board a flight to Hong Kong. For more than a month, nobody knew where he was. A month later, his wife was allowed to visit him in detention. A vague news report claimed he was facing charges of tax evasion, although the tax bureau, when contacted by journalists about the case, claimed to have no information. In mid-June 2011 he was finally released.

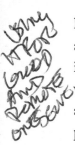

After the devastating 2008 earthquake in Sichuan province, Ai used his fame and influence to support and promote efforts by local families, activist bloggers, and human rights lawyers to compile and disseminate information about the children who had died in schools that collapsed while surrounding buildings remained intact. Corrupt local officials had allowed construction firms to cut corners on school buildings across the province. Angry parents wanted answers. They wanted those responsible to be punished. Instead the government detained several activists and convicted a few on charges of subversion. Names of people and places related to this accountability movement, as well as a range of related phrases, were put on the list of things that Internet companies must remove from blogs and social networking sites. Ai Weiwei's name, and all discussion of his case, is also banned from China's social networking platforms.

If Facebook—or any other global social network—were to set up shop in China, authorities will expect company employees to block Ai Weiwei fan pages and to erase pages created by families in Sichuan to raise awareness about what happened to their children. Police will expect the company to hand over all personally identifying information about the Chinese users who create such pages, and suspend their accounts. If activist programmers try to create iPhone apps through which people can report corruption, or a "Free Ai Weiwei" app featuring pictures of his art, they can count on being censored by Apple's Chinese app store. Such is the price of doing business in China. Similar prices are demanded with increasing frequency by governments around the world.

The first time I visited Ai Weiwei at his home was on a frosty Beijing morning in January 2009. He was just having breakfast. We sat at his kitchen table as he ate a bowl of rice porridge and a couple of meat-filled steamed buns, and talked about personal responsibility. "Either you're part of the crime, or you're compassionate," he said. "We will never have a real civil society, a democratic society, unless people take responsibility." He took a bite of steamed bun and washed it down with porridge, then continued:

> Why do I want to take any responsibility? Democracy is not a political ideal. Democracy is a means of handling problems. This method is effective—why? Because everybody in society takes responsibility. If nobody is taking responsibility, it shouldn't be called "society." Or it's a slave society anyway. . . .
>
> Even people in the police, even people who make policy, they are all able to make choices. Otherwise my blog wouldn't survive. There are always people who insist. One person says, "This post has to be deleted," but another says, "It's best not to delete it." I believe somebody must have worked to make it happen. So I believe the desire for justice and equality is something that people must have in their own hearts. This isn't something that one

person can give to another. This is a right that must be exercised. If you don't exercise your right, society will be in a difficult state.

Later that year Ai's accounts on two different Chinese blog-hosting platforms were deactivated.

It will take time—and much work and sacrifice by a great many people—to figure out how to bring "consent of the networked" to life in a messy, complicated, and rapidly evolving world. The way forward will most certainly involve a great deal of trial and error. Conflict and abuse of power will never disappear from any human society. But as Ai points out, every person has the ability to influence the political future depending on who and where we are.

Whether we are simply users of technology, investors in technology companies, employees or executives of Internet-related companies, elected officials, or mid-ranking government bureaucrats, we all have a responsibility to do whatever we can to prevent abuse of digital power, and avoid abusing it ourselves. We have a responsibility to hold the abusers of digital power to account, along with their facilitators and collaborators. If we do not, when we wake up one morning to discover that our freedoms have eroded beyond recognition, we will have only ourselves to blame.

http://consentofthenetworked.com

SHE WANTS THE
INTERNET TO BE POWERD
BY PEOPLE (US) LESS
SURVACANLE —

AFTERWORD TO
THE PAPERBACK EDITION

In late January 2012, thousands of people across Poland took to the streets to protest the Anti-Counterfeiting Trade Agreement (ACTA). The treaty had already been signed in late 2011 by European Union trade negotiators and twenty-two EU member states without much media attention, but by early February anti-ACTA protests had spread to over two hundred cities across Europe. Politicians got the message. On July 4, 2012, the European Parliament voted overwhelmingly against ratification. Several dozen parliamentarians held up bright yellow signs: HELLO DEMOCRACY, GOODBYE ACTA.

Europe's rejection of ACTA was just one victory of a global movement for digital liberty that came into its own in 2012. As Chapter 7 described it, ACTA was conceived by the United States and negotiated over the course of several years—initially in secret—with thirty-four other nations. For years, debates over ACTA—and related debates over how to balance intellectual property rights and online free-speech rights—had been confined to relatively obscure and specialized communities of activists, lawyers, and academics. That has changed, as the global netizen-rights movement to counter abuses of digital power has grown from infancy to adolescence.

I ended *Consent of the Networked* with a call for action, and in 2012 netizens around the world proved they are willing to act, as demonstrated by the movement's recent successes. But while we have gained momentum, we face continuing challenges in the pursuit of digital liberty that will not easily be overcome.

VICTORIES AND LESSONS

The movement's first victory of 2012 came in late January with the defeat in the US House of Representatives of the Stop Online Piracy Act (SOPA) and its sister bill in the Senate, the PROTECT IP Act. If passed, SOPA would have compelled social media companies to monitor and censor users, to prevent those users from violating somebody else's copyright. Failure to do so could have resulted in blacklisting of websites or prosecution of those sites' owners. Both SOPA and PROTECT IP would also have empowered the US Attorney General to order the blockage of allegedly infringing websites—based anywhere on Earth.

On January 18 Wikipedia and thousands of other websites went offline for twenty-four hours to make a dramatic point about how overzealous copyright enforcement on the Internet would result in de facto online censorship in the United States and around the world. Throughout the month, millions of Americans delivered the message through e-mails, phone calls, and letters to elected representatives that their position on SOPA or PROTECT IP could be a voting issue. By January 20, over two hundred members of Congress had declared their opposition and several dozen key supporters from both parties had changed their minds in response to voter pressure. Both bills were scrapped in one of the most rapid reversals of support for a major piece of legislation in Washington on any issue in years.

Amazingly, just three months earlier the bills' powerful Republican and Democratic sponsors—leading members of both parties in both houses of Congress—had been confident that they could garner enough bipartisan support to pass them. Intellectual property law had never been a hot media story. Copyright has always been a vital issue for the entertainment and software industries, whose donations help to fill many congressional campaign coffers, as discussed in the "Lobbynomics" section of Chapter 7, but prior to 2012, copyright had never threatened to become a serious concern for a critical mass of voters.

A broader cross section of Americans started to care about the potential free speech and privacy implications of overly broad intellectual property law only after activists joined forces with other groups scrambling to protect their interests: Internet companies (and their investors), which worried that such legislation would destroy their business; nonprofit Internet communities, such as Wikipedia, which feared untenable legal burdens; academic experts who had been trying for years to call public attention to these issues; and a group of engineers who had been involved with the Internet's creation. This broad alliance in turn rallied loosely knit coalitions of Internet users generally concerned about censorship and surveillance, who convinced their friends to sign petitions and call their elected representatives. The bills' supporters did their best to rally business associations and trade unions to their side, but the pro-SOPA "save Hollywood" argument could not compete with the grassroots appeal to "save our Internet." Harvard law professor Yochai Benkler described the events of late 2011 and January 2012 as the maturation of a "networked public sphere"—through which public discourse and activism were able to break through the usual media and political gatekeepers.

In the second half of 2012, some of the key players from the anti-SOPA and anti-ACTA battles applied similar "united front"–style tactics to the global battle over Internet governance. As Chapter 13 described, an international tug-of-war has been under way for nearly two decades over whether standard-setting, coordination, and governance of core Internet technologies, resources, and operational protocols should be managed by nongovernmental, multi-stakeholder bodies or instead by governments through the United Nations. That conflict came to a head at a meeting of the International Telecommunications Union (ITU) in Dubai in December 2012, as some governments sought to use their membership in the obscure UN body to expand their own power over coordination, rule-making, and standards-setting for the global Internet.

In the late spring and early summer, Google and other major Internet-industry players, the US government, and a number of European governments joined forces with a community of academic experts, policy wonks, and human rights groups who had been fighting uphill battles for years to advance human-rights concerns within the world's global Internet governance bodies and structures. As with the intellectual property debates, this loose alliance of players brought Internet governance issues out of obscurity and onto the front pages of newspapers, and into major TV newscasts in an unprecedented way.

On the heels of the victories against SOPA and ACTA, the ITU's plan to hold a closed-door meeting to negotiate and approve government proposals, without any formal public consultation process, appeared jarringly out of tune with the times. The two-week meeting in Dubai came under a level of global public scrutiny and criticism for which the ITU secretariat and government delegates were completely unprepared. In the end they were forced to open up their documents and meetings to the public—thanks in no small part to activists who first obtained leaked documents and other information about the negotiations and posted it all online, regardless of what the organization agreed to.

Last-ditch efforts by China, Russia, Iran, and others to transfer authority over the global domain-name system from ICANN (Internet Corporation for Assigned Names and Numbers) to the ITU, proposals to institutionalize surveillance at a global level, and initiatives to require Internet companies to pay telecommunications companies a fee when users visit their websites were all blocked—primarily by democratic governments participating in the process. In the end, however, the treaty revisions retained language referencing the Internet in ways that opponents believed amounted to vaguely defined assertions of ITU authority over aspects of Internet governance. Fifty-five countries walked away without signing, including the United States, the UK, Canada, Costa Rica, the Czech Republic, Denmark, Egypt, Kenya, the Netherlands, New Zealand, Poland, Qatar, Sweden, and many others—mainly but not exclusively democracies. With so many dissenters and other representatives holding off on signing indefinitely, the treaty had little chance of being effectively implemented.

Some activists saw the collapse of the ITU's treaty-making process as a victory against efforts by authoritarian governments to make the global Internet more compatible with the surveillance and censorship regimes that they have already set up in their own countries. Others saw it as the beginning of a new phase in protracted global Internet governance battles—the outcome of which may or may not end up being a net positive for Internet freedom. Yet while both are partly right, one thing is clear: even if a bloc of governments manages to pass an international treaty or agreement, that treaty will lack legitimacy without a critical mass of support from the global digitally networked public sphere.

For all their continued contradictions and hypocrisy, particularly regarding surveillance, the United States and a number of other democracies have made strategic decisions reflecting their awareness that their political legitimacy—and even their nations' long-term geopolitical and economic power—depends on building a positive relationship with the global commons. It is in their interest to champion governance norms and technical standards that will enable the open, globally interconnected netizen "commons" that I described in Chapter 2 to grow and thrive. It is also in the economic and trade interests of the developed Western democracy to maintain the present Internet governance system over which the UN has minimal influence.

Without civil society's globally networked activism efforts—organizing petition drives and signature campaigns, leaking internal ITU documents, and tweeting information about closed-door proceedings—it is difficult to imagine how these governments could have succeeded to the extent that they did in throwing a wrench into ITU encroachment on Internet governance. A number of the governments that voted no in December 2012 would likely have quietly voted yes in a different era for the sake of diplomatic compromise and unity. They found it impossible to do so knowing that their own citizens as well as people from all over the world were watching online in real time and would judge them accordingly. Civil society, industry, and democratic governments have discovered that while their interests are not always fully aligned, they nonetheless will continue to need each other in future global power struggles over Internet governance.

SOPA, ACTA, and ITU encroachments on the Internet were all defeated or beaten back by three-way strategic alliances among civil society defenders of the digital commons, powerful corporations (in all three cases Google was at the forefront), and champions within government. Some version of the three-way government–industry–civil society alliance model was also a factor in a range of victories against government encroachments on Internet freedom around the world in 2012, from the UK to Lebanon, the Philippines, and Pakistan.

Where these three elements are not aligned, the cause of digital liberty is much less likely to prevail. In Russia the absence of common cause between elements of civil society, business, and government contributed to the ease with which the latter passed draconian new censorship and surveillance measures in late 2012, despite the emergence in December 2011 of an Internet-driven protest movement. In China, as Chapter 3 described, a united front between industry and government, which both have an interest in perpetuating the censorship system (the government for obvious political reasons, industry because censorship keeps out foreign competitors), is one of many reasons the Chinese Communist Party is unlikely to relax that censorship anytime soon. In fact, as 2012 came to a close and a new generation of leaders moved into office, the censorship and surveillance system underwent a technical upgrade.

Furthermore, while there are many victories to celebrate in the United States, Europe, and elsewhere, democracies nonetheless face a troubling problem. Alliances among civil society, industry, and political champions have clearly worked against new threats, but similar alliances have proven much more difficult to build and sustain against the creeping, slow-motion attacks on citizens' digital freedom that started many years ago. The legal frameworks, technologies, and corporate-government relationships that facilitate opaque and unaccountable digital surveillance have grown entrenched while most of society (and their elected representatives) have failed to grasp emerging realities, let alone their long-term implications.

Rolling back established laws, regulations, and practices is much more difficult than fighting new bills or treaties before they are passed or implemented. It is much harder to suddenly demand greater transparency and public accountability in relationships between industry and government that have already become entrenched over a long period of time. The world urgently needs a strong, global multi-stakeholder coalition to fight against unaccountable surveillance—and to challenge the laws, technologies, national security norms, law enforcement practices, business practices, and social attitudes that combine to make surveillance abuses not only possible but globally pervasive. Such a coalition is critical if citizens of the democratic world hope to protect and defend their own democracies in the Internet age, and if the people of China and Iran are ever going to enjoy a free and open Internet.

SURVEILLANCE PLANET

On November 13, 2012, Google published its sixth biannual Transparency Report. With three years of data now published, a trend is clear: government surveillance of citizens via corporate networks is on the rise, in democracies as well as dictatorships. From July through December 2009, the first period for which Google reported information on government demands, the company says, it received 12,539 requests for user data from twenty-one governments. From January through June 2012, it received 20,938 requests from thirty-one governments. For every half-year reporting cycle in between, requests have steadily increased.

For more recent periods Google has also reported on the company's compliance rate. In the first half of 2012 the US government made 7,969 requests for user information and Google complied with 90 percent (this includes requests made by other governments via Mutual Assistance Treaties, although that breakdown was not available). Of Brazil's 1,566 requests, Google handed over user data for 76 percent. The South Korean government made 423 requests but Google complied with only 35 percent. Google complied with none of the Turkish government's

112 requests. (Google does not report statistics for countries in which it has received fewer than thirty requests.)

In 2012 Twitter started to follow Google's lead with its own transparency reporting system, as did LinkedIn, the file storage company Dropbox, and the California-based Internet service provider Sonic.net. As of this writing, no other Internet or telecommunications companies have made a public commitment to regular and systematic reporting of government demands, either to hand over user data or to remove content. Google's report for the first half of 2012 documents a number of fraudulent requests, on top of the hundreds of requests from governments around the world with which the company refused to comply because its lawyers determined them to be sufficiently questionable, even according to the laws of the countries where the requests were made.

The less transparent companies are about their data hand-over practices, the less incentive they have to reject or challenge illegitimate government requests, which as Google's data show happen everywhere. And yet so far no telecommunications company in any democracy has been willing to report systematically on the volume of government requests received, let alone the percentage with which it has complied.

In response to a 2012 *New York Times* report describing how law enforcement was routinely requesting consumers' cell phone records—sometimes with little judicial oversight and no consumer knowledge—Massachusetts Congressman Ed Markey, co-chair of the Congressional Bipartisan Privacy Caucus, queried nine major wireless carriers for information about the volume and scope of law enforcement requests. Their replies revealed that in 2011, law enforcement officials at all levels had made 1.3 million requests for user data including requests for geolocation information, content of text messages, and wiretaps, among others. While the carriers said that all requests came with a legal warrant or were granted in situations where individuals were in immediate danger, requests included so-called cell-tower dumps, in which carriers provide all the phone numbers of mobile phone users who connect with a cell tower during a specific period of time. As Markey's office explained, this often includes information on innocent people. He proceeded to ask the Justice Department how the government "handles, administers, and disposes of this information" and proposed a Wireless Surveillance Act that would require regular government disclosure about the volume and nature of its requests and also impose limitations on how long consumers' personal information can be held.

Evidence is also emerging that even when US law allows carriers to withhold information from authorities until consent is received from the customer, most fail to take measures that would protect users' privacy. In November 2012 the *New York Times* reported that New York police routinely subpoena call records for a phone when it is lost or stolen without first obtaining the consent of the phone's owner. Those records tend to include data from calls made not only by the thief, but also by the victim before the theft, as well as after the phone was retrieved or after the number is transferred to a new device. This information about phone calls made by innocent people is then retained in databases for use in potential future investigations. Of all the wireless carriers operating in New York, only one, Sprint Nextel, required police to obtain written consent from the victim before it would honor the subpoenas. AT&T, Verizon, T-Mobile, and Metro-PCS all complied without question. Let us hope that consumers and investors will choose to reward Sprint Nextel for standing up for user rights.

Networked communications technologies are evolving at breakneck speed. The legitimate needs of law enforcement and national defense have been so urgent that the mechanisms of oversight, transparency, and accountability necessary to keep abuse in check have failed to keep pace. Given the global reach of the technologies and companies involved, the consequences can be alarming. As they expand from their home countries into markets around the world, many US and European companies have carelessly transferred their flawed and excessively opaque law enforcement compliance practices to overseas operations including countries where free speech protections are weak or nonexistent, where judiciaries lack independence, and where the political process is not sufficiently democratic to keep abuses even vaguely within the bounds of international human rights norms. Even worse, as earlier chapters of this book have described, such companies are exporting specialized products specifically designed to help law enforcement track "criminals" to countries whose law enforcement agencies are known to include peaceful political activity and speech in their definition of "crime."

CORPORATE RESPONSIBILITY: OXYMORON OR OPPORTUNITY?

In April 2012 the Swedish investigative television program *Uppdrag Granskning* ran an expose of the Swedish-Finnish telecommunications company TeliaSonera and its operations in former Soviet states including Belarus, Uzbekistan, Kazakhstan, and Azerbaijan. The report quoted sources from within the company who described how its subsidiary in Azerbaijan, Azercell, installed devices known internally as "black boxes" that allow real-time blanket monitoring of all mobile traffic without court orders or warrants. According to the report, Azercell hosted an office for government security personnel on company premises. One TeliaSonera whistle-blower told a reporter, "The Arab Spring prompted the regimes to tighten their surveillance. . . . There's no limit to how much wiretapping is done, none at all."

Azerbaijan, a former Soviet republic bordered by Russia, Georgia, Armenia, and Iran, is among the countries that have embraced some version of "networked authoritarianism," a model pioneered by China and which I described at length in Chapter 3. Human rights groups have recently condemned the Azeri government for arrests of activists and journalists. Yet the Swedish government, a shareholder in Telia-Sonera, expressed little concern about revelations of the company's role in facilitating government repression in authoritarian states and took no meaningful action to address the underlying problem. In response to a reporter's question about TeliaSonera's operations in Belarus, where the regime conducts rampant surveillance on its opposition and where a brutal crackdown took place last year, Swedish Foreign Minister Carl Bildt replied, "In general, I think that it's good that we participate in developing telecommunications in different countries. Having a working mobile phone system in Belarus is better for the opposition than for the regime."

That may have been a reasonable assumption a few years ago, but as this book has described at length, across a range of countries police and militaries are embedding themselves ever more deeply into their nations' domestic Internet and mobile wireless infrastructure, diminishing the relative advantage that political underdogs and challengers once held with their mobile phones and home computers. For companies such as TeliaSonera not to take responsibility under such circumstances is no more acceptable than if Yahoo had refused to take ethical responsibility for its business decisions in the wake of revelations in 2005 and 2006 of the company's complicity in the arrest and conviction of Chinese dissidents—a case described at length in Chapter 9.

One of Yahoo's responses to its problems in China was to play an active role in creating the Global Network Initiative (GNI) along with Google and Microsoft, which also came under fire around the same time for collusion with Chinese state censorship. As described in Chapter 11, the GNI requires members to commit to core principles on free expression and privacy; conduct human rights impact assessments; engage with human rights groups, socially responsible investors, and academic researchers on how to address the risks they face; and participate in a three-phase assessment process that verifies whether and to what extent the companies are actually implementing the principles.

In 2012 GNI's three founding companies, Google, Microsoft, and Yahoo, successfully completed the first round of independent assessments. Independent assessors determined that the three companies had put in place company-wide policies and procedures so that staff would have the tools, knowledge, and support to identify and mitigate human rights risks. To receive a passing grade, the assessment results had to satisfy a board of directors representing all stakeholder groups including tough customers such as Human Rights Watch and Human Rights First, alongside socially responsible investors such as Calvert and Domini, which routinely challenge companies on human rights risks, as well as academics with an established research record on the role of companies in government censorship and surveillance. The second round of assessments to determine how those policies and procedures are actually being implemented was scheduled to take place in the first half of 2013. Facebook joined as an observer in the spring of 2012 and must decide one year later whether to join GNI before its one-year non-renewable observer status expires. Throughout 2011 and 2012 GNI worked to build policy alliances between industry and civil society in the defense of Internet users' free expression and privacy, bringing together activists and companies to strategize on problematic laws and regulations in a range of countries including Vietnam, Thailand, India, Russia, the United States, and the United Kingdom.

Might TeliaSonera have acted differently in Azerbaijan, Belarus, and elsewhere if it had been a GNI member? Between 2006 and 2008 a TeliaSonera executive actually participated alongside Yahoo, Google, and Microsoft in early discussions that eventually led to the formation of the GNI. However, the company decided not to join because it claimed that the GNI was too focused on the problems of Internet companies and did not fit the needs of telecommunications companies that build network infrastructure and whose relationships with governments are operationally and historically much more intertwined.

Still, to its credit TeliaSonera did not completely stick its head in the sand. It publicly acknowledged that its operations in Azerbaijan and elsewhere faced human rights risks and threats to free expression and privacy that the company needed to do a better job of anticipating and mitigating. In July 2012 it announced a partnership with the internationally respected Danish Institute for Human Rights to support the development of a new human rights impact assessment process, which would "include freedom of expression and privacy issues and be benchmarked on the United Nations Guiding Principles on Business and Human Rights." The company also helped bring together eleven of the world's largest telecommunications companies—Alcatel-Lucent, AT&T, BT, France Telecom-Orange, Millicom, Nokia Siemens Networks, Tele2, Telefónica, TeliaSonera, Telenor, and Vodafone—in a series of meetings on free expression and privacy problems faced by the sector. The meetings became known as the Industry Dialogue on Telecommunications, Privacy, and Freedom of Expression. (Deutsche Telecom, which owns T-Mobile, was an early participant but later dropped out.) As of January 2013 the Industry Dialogue was in discussions with the GNI about a potential partnership that if consummated would challenge its member companies to strengthen their policy and operational commitments.

Regardless of what commitments these companies finally make, their efforts will not be credible until they engage directly and systematically with a range of stakeholders including activist communities from the countries in which they operate, as well as international NGOs and researchers who specialize in free expression and privacy issues. It is also vital that they be willing to submit to an independent assessment process with sufficient rigor that their claims about commitments, policies, and practices can actually be verified. Right now the GNI, despite imperfections and growing pains, is the only organization with any such processes in place. Whether and how the telecommunications industry chooses to act has real implications, and not only for the lives of individual dissidents, activists, and investigative journalists around the world. Whether the industry takes responsibility for its human rights obligations will affect the future of democracy, accountable governance, and social justice all over the world. In the short-to-medium term the industry is already having an impact on political power dynamics in the Middle East and Africa in the wake of the events of early 2011 that the world came to know as the Arab Spring.

JUSTICE IN THE MARKETPLACE

Across much of the Middle East and Africa, struggles for accountable governance and social justice came under renewed attack in 2012. It is clear that the Internet, mobile telephony, and social media will play an important role in that region's political future. According to a 2012 Pew Research Center survey conducted across twenty countries, respondents in Arab countries were approximately twice as likely to use social media to discuss politics, community issues, and religion as respondents in the rest of the world. Less than half of Americans surveyed used social media to discuss politics and community issues. This disparity is not surprising given the array of mass media outlets and community organizations that exist in democracies. But in five Islamic countries—Turkey, Egypt, Lebanon, Jordan, and Tunisia—strong majorities relied on social media to discuss their civic and political concerns.

If the Sovereigns of Cyberspace—all of the companies whose products, platforms, and services increasingly mediate citizens' relationships with their communities and governments—cannot do right by the world's most politically vulnerable and exposed Internet users, the consequences for the entire human race will be, as programmers like to put it, nontrivial.

The movement to build and promote more secure, noncommercial communications tools, and train activists how to protect themselves when using commercial services, expanded dramatically in 2012 with Western governments and several companies providing funding to activist groups. But in a world where most Internet and mobile users do not consider themselves activists until they suddenly find themselves to be threatened or in trouble, those admirable efforts will never be adequate on their own.

In October 2012 in Brussels, the London-based activist group Privacy International and Google hosted a meeting of technologists and legal experts to draft what they decided to call the International Principles on Communications Surveillance and Human Rights. The goal, according to Privacy International, was "to provide civil society, industry and government with a framework against which to evaluate whether current or proposed surveillance laws and practices are consistent with human rights." They posted the principles online at http://necessaryandproportionate.net, launched a public consultation inviting feedback from all interested parties around the world, and organized an additional workshop with the Electronic Frontier Foundation and Latin American activist groups in December 2012.

The content of the document remains in flux as of this writing, but the principles are an attempt to lay down some markers that activists and companies can use to build common ground and common strategies when negotiating with lawmakers and regulators around the world, assuming that there is a shared goal of safeguarding Internet users' rights to free expression and privacy. They included criteria such as legality, legitimacy of purpose, necessity, proportionality, competence of the enforcement authorities, due process, user notification, transparency, and adequate oversight against abuse.

As of January 2013 only a few NGOs had signed on, and Google itself had yet to officially endorse the principles. It remained unclear whether any other companies would dare to engage with the process, let alone be prepared to stand publicly behind the final outcome that would necessarily be the product of deliberation among stakeholders from around the world. Whatever happens with the principles themselves, the effort behind them demonstrates why transparency by Internet and telecommunications companies is so important. If not for Google's Transparency Report, the public would be in the dark about the steady upward trend in worldwide government demands for Google's user data. The world's netizens have a right to know whether other Internet and telecommunications companies that they depend on are experiencing a similar increase in government demands.

Of course, those government-focused surveillance principles do nothing to address a related problem—discussed at length in several chapters of this book—that exacerbates the consequences of government surveillance. While some are doing much better than others, no company is taking adequate responsibility for how their private terms of service, privacy policies, and identity requirements ultimately affect users' political liberties in the physical world.

In the United States, activists, companies, and politicians who joined forces to defeat SOPA in early 2012 are fiercely divided over the appropriate role of government regulation in protecting Internet users' privacy. Consumer privacy issues naturally overlap with issues of government surveillance insofar as data once collected for commercial purposes can potentially be accessed, and sometimes even purchased from data brokers on the open market, by authorities who then use it for other distinctly noncommercial, law enforcement, and national security purposes.

Europe has become the battleground over the extent to which governments should be allowed to regulate Internet companies to protect citizens from abuses or negligence by those companies. Activists seeking to force Facebook to abide by European privacy laws argued that the changes they sought would protect not only European Internet users, but social media users in the Middle East who rely heavily on Facebook to inform and mobilize their communities. In December a German data protection commissioner operating in the northern state of Schleswig-Holstein ruled that Facebook's real-name policy violates the German Telemedia Act, which requires service providers to allow users to communicate and conduct transactions anonymously.

As 2012 came to an end, regulators in both Europe and the United States were also considering whether to punish Google for allegedly abusing its market dominance to manipulate search results in favor of its own products and services, at the expense of its rivals. Free-market champions, such as the *Economist,* argued strongly against increased government regulation of the world's Internet giants, on the grounds that fierce market competition among Google, Apple, Facebook, and Amazon would ultimately constrain companies from abusing Internet users' rights because the market would punish them for doing so. Certainly regulation is problematic, not least because it cannot keep up with technological and social change and it is often subject to capture by special-interest lobbies. That said, the market constrains bad behavior by companies only when consumers and investors are adequately informed about what companies are doing—both internally and in relationship to governments. The market currently lacks such information.

If all telecommunications companies running broadband and wireless services, plus Facebook and other social networking platforms, were to join Twitter and Google in releasing country-by-country data about government censorship and surveillance demands they receive, we would have made the first real step toward having an informed marketplace. Perhaps then the global commons would wake up to what is truly happening, and act as they did with the ITU, ACTA, and SOPA.

Citizens of democracies, companies that understand that they can build long-term global value for their brand by earning trust with their users, and politicians who understand the need to protect and strengthen the digital commons (even if mainly out of self-interest) must unite to demand a national and global reconsideration of already deeply entrenched surveillance laws, technologies, and corporate practices. There needs to be a more robust public debate about the facts of digital surveillance in democracies, the implications for accountable governance and social justice—and what can be done now that the surveillance state has already been allowed to reach too far, too fast. That debate requires data and information that companies as well as democratic governments have so far been reluctant to share. Companies claiming to support a free and open Internet and that benefit from the existence of a robust global digital commons are doing the commons no favors unless and until they agree to publish systematic and useable information about their relationships with governments.

Furthermore, governments that want their citizens to believe that their support for global Internet freedom and citizens' digital rights is genuine—and not shallow political rhetoric—must make sure that laws are not preventing companies from releasing such information. All governments that intend to keep calling themselves "democratic" with a straight face should publish their own transparency reports so that engaged citizens can see enough of the whole picture that they can grant or withdraw consent for, or divest from, the surveillance systems and procedures governments and companies have built. Until these things happen, Western democracies and Western companies will remain net exporters of surveillance technologies, legal norms, and business practices that facilitate political abuse of surveillance powers by repressive regimes—and that will ultimately corrode existing democracies.

Getting governments and companies to do these things will make the fights against ACTA, SOPA, and ITU Internet incursion look trivial in comparison. The global movement for digital liberty spread its wings and took flight in 2012, but the real tests of its strength and agility have yet to come.

SHE WANTS THE INTERNET
TO BE POWERD BY PEOPLE (US)
LESS SURVALANCE —

NOTES

CHAPTER 1: CONSENT AND SOVEREIGNTY

5 "Before things were simple: you had the good guys on one side, and the bad guys on the other. Today, things are more subtle": Jailan Zayan, "Egypt, Tunisia Finding that Road to Freedom Is Rocky," Agence France Presse, May 26, 2011.

6 President Barack Obama waxed enthusiastic about the political power of social networking: Full transcript at www.whitehouse.gov/the-press -office/2011/04/20/remarks-president-facebook-town-hall (accessed June 21, 2011).

7 A classic example was Google's clash with the Chinese government: A full account of those events can be found in Steven Levy, *In the Plex: How Google Thinks, Works, and Shapes Our Lives* (New York: Simon & Schuster, 2011). Also see John Pomfret, "In China, Google Users Worry They May Lose an Engine of Progress," *Washington Post*, March 20, 2010, www.washingtonpost .com/wp-dyn/content/article/2010/03/19/AR2010031900986.html (accessed June 21, 2011).

9 geopolitical vision for a digitally networked world: Eric Schmidt and Jared Cohen, "The Digital Disruption: Connectivity and the Diffusion of Power," *Foreign Affairs* 89, no. 6 (November/December 2010), 75–85.

10 In his book *The Filter Bubble*, Eli Pariser: Eli Pariser, *The Filter Bubble: What the Internet Is Hiding from You* (New York: Penguin Press, 2011).

10 Siva Vaidhyanathan warns: Siva Vaidhyanathan, *The Googlization of Every-thing (And Why We Should Worry)* (Berkeley: University of California Press, 2011).

10 As Harvard's Joseph Nye points out in *The Future of Power*: Joseph S. Nye Jr., *The Future of Power* (New York: PublicAffairs, 2011).

11 Other kinds of transnational organizations are also challenging the power of nation-states: One of the earliest policy analyses of the challenge posed by transnational organizations to the power of nation-states, and how the Internet had amplified the power of new actors, was by Jessica T. Matthews, "Power Shift," *Foreign Affairs* 76, no. 1 (January/February 1997), 50–66. Parag Khanna argues that the world is entering a new phase that he calls the

"new medievalism," in which states must share power with other transnational actors. Parag Khanna, *How to Run the World: Charting a Course to the Next Renaissance* (New York: Random House, 2011).

13   *Communication Power*: Manuel Castells, *Communication Power* (New York: Oxford University Press, 2009), 346–362, 431–432.

CHAPTER 2: RISE OF THE DIGITAL COMMONS

17   **In his book *The Wealth of Networks*:** Yochai Benkler, *The Wealth of Networks* (New Haven, CT: Yale University Press, 2007), 472.

18   **The World Wide Web, invented two decades after the Internet:** For Berners-Lee's firsthand account, see Tim Berners-Lee, *Weaving the Web: The Original Design and Ultimate Destiny of the World Wide Web* (New York: HarperOne, 1999).

20   **In 1989, the computer scientist Richard Stallman got the ball rolling:** Though the distinction may seem arcane to nonprogrammers, there are important philosophical differences between developers of open-source software—a much broader set of people—and adherents of the free software movement. As Stallman puts it, "Open source is a development methodology; free software is a social movement." See Richard Stallman, "Why Open Source Misses the Point of Free Software," www.gnu.org/philosophy/open -source-misses-the-point.html. Also see the Linux Information Project's definition of open-source software at www.linfo.org/open_source.html.

20   **By 2008, 60 percent of the world's computer servers ran on Linux, with a minority 40 percent using Microsoft's proprietary Windows platform:** James Niccolai, "Ballmer Still Searching for an Answer to Google," *IDG News*, September 26, 2008, www.pcworld.com/businesscenter/article/151568/ballmer _still_searching_for_an_answer_to_google.html (accessed June 21, 2011).

21   **since 2005 around 6,100 individual developers and six hundred companies have contributed to the Linux kernel:** Jonathan Corbet et al., "Linux Kernel Development: How Fast It Is Going, Who Is Doing It, What They Are Doing, and Who Is Sponsoring It," Linux Foundation, December 2010, www.linuxfoundation.org/docs/lf_linux_kernel_development_2010.pdf.

21   **In a 2009 speech, Jim Zemlin, executive director of the Linux Foundation, pointed out that every Internet user on the planet is a Linux user in some way:** Steven J. Vaughan-Nichols, "How Many Linux Users Are There (Really)?" *LinuxPlanet*, February 18, 2009, www.linuxplanet.com/linuxplanet /reports/6671/1.

23   **CC licenses encourage people to copy and republish activist content:** See http://creativecommons.org.

24   **human rights groups, and journalists who have since gone back and analyzed the Tunisian revolution:** See, for example, Mark Lynch, Blake Hounshell, and Susan Glasser, eds., *Revolution in the Arab World: Tunisia, Egypt, and the Unmaking of an Era* (Washington, DC: Foreign Policy Institute, 2011).

24 Clay Shirky, a technology and social theorist at New York University: Clay Shirky, *Here Comes Everybody: The Power of Organizing Without Organizations* (New York: Penguin Press, 2008); and Clay Shirky, *Cognitive Surplus: Creativity and Generosity in a Connected Age* (New York: Penguin Press, 2010).

25 Similar tools and activities were key to the Tea Party–led Republican takeover of the US House of Representatives in 2010: For example, see Matt Bai, "D.I.Y. Populism, Left and Right," *New York Times*, October 30, 2010, www.nytimes.com/2010/10/31/weekinreview/31bai.html (accessed June 21, 2011).

25 As Harvard law professor Lawrence Lessig explained more than a decade ago in his seminal book, *Code and Other Laws of Cyberspace*: Lawrence Lessig, *Code and Other Laws of Cyberspace* (New York: Basic Books, 1999).

26 "netizenship": See David R. Johnson, "Democracy in Cyberspace: Self-Governing Netizens and a New, Global Form of Civic Virtue, Online," in *The Next Digital Decade: Essays on the Future of the Internet*, ed. Berin Szoka and Adam Marcus (Washington, DC: Techfreedom, 2010).

26 David Bollier, whose book *Viral Spiral* documents the rise of the digital commons in the United States: David Bollier, *Viral Spiral: How the Commoners Built a Digital Republic of Their Own* (New York: New Press, 2009).

CHAPTER 3: NETWORKED AUTHORITARIANISM

32 Peking University: Though the city's English name is spelled "Beijing," the university, founded in 1898, has opted to keep its prerevolutionary English spelling.

32 what I call China's networked authoritarianism: For a more academic analysis of this phenomenon, see Rebecca MacKinnon, "China's 'Networked Authoritarianism,'" *Journal of Democracy* 22, no. 2 (April 2011): 34–46, http://new america.net/publications/articles/2011/china_s_networked_authoritarianism _49202. Also see Rebecca MacKinnon, "China's Censorship 2.0: How Companies Censor Bloggers," *First Monday* 14, no. 2 (January 25, 2009), http://firstmonday.org/htbin/cgiwrap/bin/ojs/index.php/fm/article/view/2378/2089.

33 In 2005 a PBS documentary crew visited university campuses around Beijing: www.pbs.org/wgbh/pages/frontline/tankman.

34 According to the Dui Hua Foundation: "Chinese State Security Arrests, Indictments Doubled in 2008," *Dui Hua Human Rights Journal*, March 25, 2009, www.duihua.org/hrjournal/2009/03/chinese-state-security-arrests.html (accessed June 27, 2011).

34 The communications revolution has transformed China in many ways: The best academic overview of the Internet's effect on Chinese society and political culture is Guobin Yang, *The Power of the Internet in China: Citizen Activism Online* (New York: Columbia University Press, 2009).

38 The government's State Council Information Office issued a direct and detailed order on the subject to all websites and news organizations: Xiao

Qiang, "The Ministry of Truth Limits Reporting on Google in China," *China Digital Times,* March 23, 2010, http://chinadigitaltimes.net/2010 /03/the-latest-directives-from-the-ministry-of-truth-032310 (accessed June 27, 2011).

39 **350 million pieces of "harmful content" had been deleted from the Chinese Internet over the course of one year:** "China Shuts Over 60,000 Porn Websites This Year," Reuters, December 30, 2010. For the full text of Wang's leaked speech about the government's Internet control strategy, see Wang Chen, "Concerning the Development and Administration of Our Country's Internet," *China Rights Forum,* 2010, no. 2, published by Human Rights in China, http://www.hrichina.org/crf/article/3241.

41 **The people who rallied to free Guo did not spring immaculately from cyberspace:** Between 2005 and late 2009 a community of internationalist, liberal-minded Chinese bloggers, web developers, and entrepreneurs held annual blogger conferences in different cities around China. I attended four of the five conferences. My account of the 2009 meeting can be found at Rebecca MacKinnon, "Chinese BloggerCon 2009: Micro Power from the Mouth of a Cave," RConversation blog, November 13, 2009, http://rconversation .blogs.com/rconversation/2009/11/chinese-bloggercon-2009-micro-power .html. Accounts of previous gatherings can be found at http://rconversation .blogs.com/rconversation/cnbloggercon.

42 **Min Jiang, a scholar of the Chinese Internet at the University of North Carolina at Charlotte, calls this sort of lively but limited public discourse "authoritarian deliberation":** Min Jiang, "Authoritarian Deliberation on Chinese Internet," *Electronic Journal of Communication* 20 (2010), http://papers .ssrn.com/sol3/papers.cfm?abstract_ id=1439354. Other academic works discuss the limits to political liberation on the Chinese Internet. Hong Kong–based scholar Jack Linchuan Qiu describes how "working-class ICTs" (Internet and communications technologies) provide new levers for government and corporations to organize and control a new class of "programmable labor." He concludes that "working-class ICTs by themselves do not constitute a sufficient condition for cultural and political empowerment." Jack Linchuan Qiu, *Working-Class Network Society: Communication Technology and the Information Have-Less in Urban China* (Cambridge, MA: MIT Press, 2009). Singapore-based scholar Yongnian Zheng describes the difference between two categories of online activism: "voice" activism (which is tolerated and sometimes encouraged because it actually helps the central government to reduce local corruption and improve governance), and "exit" activism (activism centered around regime change, which is not tolerated). See Yongnian Zheng, *Technological Empowerment: The Internet, State, and Society in China* (Stanford, CA: Stanford University Press, 2008), 164–165.

43 **President Hu Jintao, who gave his blessing to the whole enterprise:** "Hu Jintao Talks to Netizens via People's Daily Online," *People's Daily Online,* June 20, 2008, http://english.people.com.cn/90001/90776/90785/6433952.html (accessed June 27, 2011).

44   "fifty-cent party": See David Bandurski, "China's Guerrilla War for the Web,"
     Far Eastern Economic Review 171, no. 6 (July 2008): 41–44, http://feer.wsj
     .com/essays/2008/august/chinas-guerrilla-war-for-the-web (accessed June
     27, 2011).

46   researchers affiliated with the University of Toronto uncovered a cyber-
     espionage network: "Tracking GhostNet: Investigating a Cyber Espionage
     Network," *Information Warfare Monitor*, March 29, 2009, www.scribd
     .com/doc/13731776/Tracking-GhostNet-Investigating-a-Cyber-Espionage
     -Network.

46   In a 2007 book about China's patriotic hacker communities: Scott J. Hen-
     derson, *The Dark Visitor: Inside the World of Chinese Hackers* (Morrisville, NC:
     Lulu Press, 2007). Also see the author's website, www.thedarkvisitor.com.

47   Capitalist investors are actually helping the Chinese Communist Party
     strengthen and refine a gilded cage: See Bill Bishop, "China's Internet: The
     Invisible Birdcage," *China Economic Quarterly* 14, no. 3 (September 2010):
     26–30; text available online at http://billbishop.tumblr.com/post/340300
     7567/chinas-internet-the-invisible-birdcage (accessed June 27, 2011).

48   University and corporate networks suffered disruptions as "punishment":
     Oiwan Lam, "China: Cracking Down Circumvention Tools," Global Voices
     Advocacy, May 13, 2011, http://advocacy.globalvoicesonline.org/2011/05
     /13/china-cracking-down-circumvention-tools.

48   To access any other websites or services from outside of China, students are
     charged: Erica Newland, "At Chinese Universities There's a Fee for 'Free,'"
     Center for Democracy and Technology, November 18, 2010, http://
     cdt.org/blogs/erica-newland/chinese-universities-theres-fee-free.

49   According to the Boston Consulting Group, China's e-commerce market is
     expected to soon become the world's largest: "Another Digital Gold Rush,"
     The Economist, May 12, 2011, www.economist.com/node/18680048?story
     _id=18680048.

49   US Secretary of State Warren Christopher reinforced this assumption: War-
     ren Christopher, *In the Stream of History: Shaping Foreign Policy for a New
     Era* (Palo Alto, CA: Stanford University Press, 1998), 159.

CHAPTER 4: VARIANTS AND PERMUTATIONS

51   When the Egyptian government shut down the Internet: Initially one
     Egyptian ISP, Noor, remained online for a few days but was eventually shut
     down as well. For a detailed technical analysis of the shutdown, see James
     Cowie, "Egypt Leaves the Internet," Renesys Blog, January 28, 2011,
     www.renesys.com/blog/2011/01/egypt-leaves-the-internet.shtml; and Earl
     Zmijewski, "Egypt's Net on Life Support," Renesys Blog, January 31, 2011,
     www.renesys.com/blog/2011/01/egypts-net-on-life-support.shtml. Also see
     James Glanz and John Markoff, "Egypt Leaders Found 'Off' Switch for In-
     ternet," *New York Times*, February 15, 2011, www.nytimes.com/2011
     /02/16/technology/16internet.html (accessed June 27, 2011).

52   **After Muammar Gaddafi cut off phone and Internet service to rebel-held areas in eastern Libya:** Margaret Corker and Charles Levinson, "Rebels Hijack Gadhafi's Phone Network," *Wall Street Journal*, April 13, 2011, http://online.wsj.com/article/SB10001424052748703841904576256512991215284.html (accessed June 27, 2011).

53   **In a comprehensive book about technology and politics in the Islamic world:** Philip N. Howard, *The Digital Origins of Dictatorship and Democracy: Information Technology and Political Islam* (New York: Oxford University Press, 2010), 10.

54   **Ahmadinejad is not the most conservative player:** For an analysis of Ahmadinejad and Iran's power struggles, see Massoumeh Torfeh, "Ahmadinejad Has Fuelled Iran's Power Struggle," *The Guardian*, May 21, 2011, www.guardian.co.uk/commentisfree/2011/may/21/mahmoud-ahmadinejad-iran-power-struggle. Also see J. David Goodman, "Iran Rift Deepens with Arrest of President's Ally," *New York Times*, June 23, 2011, www.nytimes.com/2011/06/24/world/middleeast/24iran.html (accessed June 27, 2011).

54   **As the anonymous editor of the underground e-mail newspaper *Kyaboon* told Global Voices:** Fred Petrossian, "Iran: Protests Prompt Emergence of Underground Internet Newspapers," Global Voices Online, July 16, 2009, http://globalvoicesonline.org/2009/07/16/iran-protests-prompt-emergence-of-underground-internet-newspapers.

54   **Iran overtook China in 2009 as the world's top jailer of journalists and bloggers, then tied with China in 2010:** See *Attacks on the Press in 2010: A Worldwide Survey by the Committee to Protect Journalists*, 2011, www.cpj.org/2011/02/attacks-on-the-press-2010.php; and *Attacks on the Press 2009: A Worldwide Survey by the Committee to Protect Journalists*, 2010, http://cpj.org/2010/02/attacks-on-the-press-2009.php.

55   **transcripts of text messaging sessions during interrogation:** *Internet Filtering in Iran: 2009*, OpenNet Initiative, June 16, 2009, http://opennet.net/research/profiles/iran.

55   **people detained in 2009 reported having the contents of their own e-mails as well as other people's e-mails read to them during interrogation:** Sanja Kelly and Sarah Cook, eds., *Freedom on the Net 2011: A Global Assessment of Internet and Digital Media Freedom*, Freedom House, April 18, 2011, downloadable at www.freedomhouse.org/report/freedom-net/freedom-net-2011. Also see Farnaz Fassihi, "Iranian Crackdown Goes Global," *Wall Street Journal*, December 3, 2009, http://online.wsj.com/article/SB125978649644673331.html (accessed June 27, 2011).

55   **In 2011 the Iranian government announced a move to the next level:** "Government Develops 'National Internet' to Combat International Internet's Impact," Reporters Without Borders, August 3, 2011, http://en.rsf.org/iran-government-develops-national-03-08-2011,40738.html (accessed August 3, 2011).

55   **Another ministry official was quoted telling Iran's news agency:** Christopher Rhoads and Farnaz Fassihi, "Iran Vows to Unplug Internet," *Wall Street Jour-*

nal, May 28, 2011, http://online.wsj.com/article/SB1000142405274870
4889404576277391449002016.html (accessed August 3, 2011).

57 **"Mobiles are too closed, the mobile operators too vulnerable":** Cory Doctorow, "Report: Belarusian Mobile Operators Gave Police List of Demonstrators," BoingBoing.net, January 15, 2011, www.boingboing.net/2011
/01/15/report-belarusian-mo.html; and Hans Rosen, "Ericsson Technology
Used to Wiretap in Belarus," *Dagens Nyheter,* December 22, 2010, www.dn.se
/nyheter/varlden/ericsson-technology-used-to-wiretap-in-belarus (all accessed August 13, 2011).

57 **text messaging services went down roughly nine hours before Iran's presidential election:** Rob Faris and Rebekah Heacock, "Cracking Down on Digital Communication and Political Organizing in Iran," OpenNet Initiative,
June 15, 2009, http://opennet.net/blog/2009/06/cracking-down-digital
-communication-and-political-organizing-iran.

58 **In 2008, a group of tech-savvy activists conducted tests that proved the
Tunisian government was already using DPI:** Sami Ben Gharbia, "Silencing
Online Speech in Tunisia," Global Voices Advocacy, August 20, 2008,
http://advocacy.globalvoicesonline.org/2008/08/20/silencing-online-speech
-in-tunisia (accessed June 27, 2011).

58 **the Iranian government had also deployed DPI technology:** See "Update on
Internet censorship in Iran," Tor Blog, January 20, 2011, https://blog.torproject
.org/blog/update-internet-censorship-iran; and "New Blocking Activity
from Iran," Tor Blog, January 9, 2011, https://blog.torproject.org/blog/new
-blocking-activity-iran.

59 **DPI is an invention of companies based in the democratic West:** See
Christopher Rhoads and Loretta Chao, "Iran's Web Spying Aided by Western Technology," *Wall Street Journal,* June 22, 2009, http://online.wsj.com
/article/SB124562668777335653.html (accessed June 27, 2011). For more
on DPI and its origins, see Hal Abelson, Ken Ledeen, and Chris Lewis, "Just
Deliver the Packets," in *Essays on Deep Packet Inspection,* Office of the Privacy
Commissioner of Canada, 2009, http://dpi.priv.gc.ca/index.php/essays/just
-deliver-the-packets; and Ralf Bendrath, "Global Technology Trends and
National Regulation: Explaining Variation in the Governance of Deep
Packet Inspection," paper presented at the International Studies Annual
Convention, New York City, February 2009, International Studies Associa-
tion, http://userpage.fu-berlin.de/~bendrath/Paper_Ralf-Bendrath_DPI
_v1-5.pdf.

59 **the Egyptian government had purchased DPI technology from a company
called Narus:** Timothy Karr, "One US Corporation's Role in Egypt's Brutal
Crackdown," *Huffington Post,* January 28, 2011, www.huffingtonpost.com
/timothy-karr/one-us-corporations-role-_b_815281.html (accessed June 27,
2011).

59 **Narus signed a multimillion-dollar deal in 2005 with Giza Systems of
Egypt:** Trevor Lloyd-Jones, "Narus Signs Regional Licence with Giza Systems," *Business Intelligence Middle East,* September 14, 2005, www.bi-me

.com/main.php?id=2047&t=1 (accessed June 27, 2011). See also p. 10 of a Giza Systems newsletter, mentioning the licensing of Narus technology to Saudi Arabia and Libya: www.gizasystems.com/admin%5CNewsLetter %5CPDF%5C8.pdf.

59 **Karr's Free Press and the Paris-based Reporters Without Borders also lodged inquiries with Cisco Systems:** Timothy Karr and Clothilde Le Coz, "Corporations and the Arab Net Crackdown," March 25, 2011, www.fpif.org /articles/corporations_and_the_arab_net_crackdown (accessed June 27, 2011).

60 **a business offer for an intrusion and surveillance software product called FinFisher:** Eli Lake, "British Firm Offered Spy Software to Egypt," *Washington Times*, April 25, 2011, www.washingtontimes.com/news/2011/apr /25/british-firm-offered-spy-software-to-egypt; Hussein uploaded the original document online: http://moftasa.posterous.com/finfisher-intrusion -software-spy-on-email-sof. Also see Ramy Raouf, "Egypt: How Companies Help the Government Spy on Activists," Global Voices Advocacy, May 7, 2011, http://advocacy.globalvoicesonline.org/2011/05/07/egypt-how-companies -help-the-government-spy-on-activists (all accessed June 27, 2011).

60 **censorship software developed and sold by North American companies is being used to block social and political content in Bahrain:** Helmi Noman and Jillian York, *West Censoring East: The Use of Western Technologies by Middle East Censors, 2010–2011*, OpenNet Initiative, March 2011, http://opennet.net/sites/opennet.net/files/ONI_WestCensoringEast.pdf. Also see Paul Sonne and Steve Stecklow, "US Products Help Block Mideast Web," *Wall Street Journal*, March 28, 2011, http://online.wsj.com/article /SB10001424052748704438104576219190417124226.html (accessed June 27, 2011).

62 **government supporters launched aggressive efforts to discredit domestic and international critics:** See, for example, Bhumika Ghimire, "Bahrain: Pro-Government Activists Are Blogging Too," Global Voices Advocacy, April 18, 2011, http://advocacy.globalvoicesonline.org/2011/04/18/bahrainpro -government-activists-are-blogging-too. Meanwhile, Ali Abdulemam—still in hiding—was sentenced in absentia to fifteen years in prison: Leila Nachawati, "Bahrain: Leading Blogger Ali Abdulemam Sentenced to 15 Years in Prison, Along with Other Human Rights Defenders," Global Voices Advocacy, June 22, 2011, http://advocacy.globalvoicesonline.org/2011/06 /22/bahrain-leading-blogger-ali-abdulemam-sentenced-to-15-years-in -prison-along-with-other-human-rights-defenders (all accessed June 27, 2011).

63 **statement by King Hamad bin Isa Al Khalifa:** "His Majesty Stresses the Key to Reform Is Through Press Freedom," Bahrain News Agency, May 3, 2011, www.bna.bh/portal/en/news/455101 (accessed August 11, 2011).

64 **In Syria, where between March and July 2011 an estimated 1,400 people were killed and at least 15,000 detained:** See Neil MacFarquhar and Rick

Gladstone, "Outside Pressure Builds on Syria," *New York Times*, August 2, 2011, www.nytimes.com/2011/08/03/world/middleeast/03syria.html; and "Syria: Mass Arrest Campaign Intensifies," Human Rights Watch, July 20, 2011, www.hrw.org/news/2011/07/20/syria-mass-arrest-campaign -intensifies (all accessed August 2, 2011).

64 "man in the middle" attack on Syrian Facebook users: See Anas Qtiesh, "Did Syria Replace Facebook's Security Certificate with a Forged One?" Global Voices Advocacy, May 4, 2011, http://advocacy.globalvoicesonline.org /2011/05/05/did-syria-replace-facebooks-security-certificate-with-a-forged -one; and Leila Nachawati, "Syrian Uprisings and Official vs. Decentralized Communications," April 27, 2011, http://advocacy.globalvoicesonline.org /2011/04/27/syrian-uprisings-and-official-vs-decentralized-communications (all accessed June 27, 2011).

64 In May, an organization called the Syrian Electronic Army (SEA) emerged: Helmi Noman, "The Emergence of Open and Organized Pro-Government Cyber Attacks in the Middle East: The Case of the Syrian Electronic Army," Information Warfare Monitor, May 30, 2011, www.infowar-monitor .net/2011/05/7349.

65 In June, Assad praised SEA directly: Danny O'Brien, "Syria's Assad Gives Tacit OK to Online Attacks on Press," June 24, 2011, www.cpj.org/internet /2011/06/syrias-assad-gives-tacit-ok-to-online-attacks-on-p.php#more.

65 the SEA claimed responsibility for attacking the website of the French embassy: "Syrian Electronic Army: Disruptive Attacks and Hyped Targets," Information Warfare Monitor, June 25, 2011, www.infowar-monitor.net /2011/06/syrian-electronic-army-disruptive-attacks-and-hyped-targets.

67 what researchers at the OpenNet Initiative call second- and third-generation Internet controls: Ronald J. Deibert, John G. Palfrey, Rafal Rohozinski, and Jonathan Zittrain, eds., *Access Controlled: The Shaping of Power, Rights, and Rule in Cyberspace* (Cambridge, MA: MIT Press, 2010).

68 One example is the use of cyber-attacks to sideline the website of jailed Russian oligarch Mikhail Khodorkovsky: "Khodorkovsky's Website Attacked amid Announcement of Sentencing," RIA Novosti, December 27, 2010, http://en.rian.ru/russia/20101227/161951482.html (accessed June 27, 2011).

68 "informal requests to companies for removal of information": Ronald Deibert and Rafal Rohozinski, "Control and Subversion in Russian Cyber-space," in *Access Controlled*, ed. Ronald J. Deibert et al. (Cambridge, MA: MIT Press, 2010), 15–34.

69 One of Russia's most popular political bloggers, Alexey Navalny: See Julia Ioffe, "Net Impact: One Man's Cyber-crusade Against Russian Corruption," *New Yorker*, April 4, 2011.

69 As Navalny told *Global Voices* contributor Gregory Asmolov: See Gregory Asmolov, "Russia: Blogger Navalny Tries to Prove that Fighting Regime Is Fun," Global Voices Online, October 27, 2010, http://globalvoicesonline

.org/2010/10/27/russia-blogger-alexey-navalny-on-fighting-regime (accessed June 27, 2011).

69   **people who had made donations to his project began to receive mysterious and threatening phone calls:** Ashley Cleek, "Russia: Anti-Corruption Donor Details Leaked," Global Voices Online, May 4, 2011, http://globalvoices online.org/2011/05/04/russia-anti-corruption-donor-details-leaked (accessed June 27, 2011).

CHAPTER 5: ERODING ACCOUNTABILITY

75   **Cybersecurity and Internet Freedom Act of 2011:** Full text can be found at http://hsgac.senate.gov/download/s-413-bill-text (accessed June 21, 2011).

75   **Lieberman told CNN, "Right now, China, the government, can disconnect parts of its Internet in a case of war. We need to have that here, too":** Chloe Albanesius, "Lieberman Backs Away from 'Internet Kill Switch,'" *PCMag*, June 21, 2010, www.pcmag.com/article2/0,2817,2365393,00.asp (accessed June 21, 2011).

75   **"precise and targeted authorities to the president":** See explanation of the bill on Lieberman's website, "Lieberman, Collins, Carper Introduce Bill to Address Serious Cyber Security Threats," February 18, 2011, http://lieberman .senate.gov/index.cfm/news-events/news/2011/2/lieberman-collins-carper -introduce-bill-to-address-serious-cyber-security-threats (accessed June 21, 2011).

76   **Mark Klein, a technician who retired from his job with AT&T in San Francisco in 2004:** For one account of Klein's discovery, see Tim Wu, *The Master Switch: The Rise and Fall of Information Empires* (New York: Random House, 2010), 249–251. A full statement by Klein can be downloaded at www.eff .org/cases/att/attachments/unredacted-klein-declaration.

77   **The Obama administration surprised many liberal voters:** The "Get Fisa Right" activist group maintains a website at http://get-fisa-right.wetpaint .com. Also see Linda Feldmann, "Left Lacks Leverage to Stop Obama's Rightward Tack," *Christian Science Monitor*, July 4, 2008, www.csmonitor.com /USA/Politics/2008/0703/p25s23-uspo.html (accessed June 27, 2011).

78   **The goal of the coalition, called Digital Due Process (DDP):** See http:// digitaldueprocess.org; also see Declan McCullagh, "Senator Renews Pledge to Update Digital-Privacy Law," *CNet*, June 16, 2011, http://news.cnet.com /8301-31921_3-20071670-281/senator-renews-pledge-to-update-digital -privacy-law (accessed June 27, 2011).

78   **At a May 2011 congressional hearing on mobile data collection and privacy:** Kashmir Hill, "DOJ to Senators During Mobile Privacy Hearing: 'We Want MORE Data,'" Forbes.com, May 10, 2011, http://blogs.forbes.com /kashmirhill/2011/05/10/doj-to-senators-during-mobile-privacy-hearing -we-want-more-data; the link to the full hearing video and testimonies can

be found at http://judiciary.senate.gov/hearings/hearing.cfm?id=e655f9e2
809e5476862f735da16bd1e7 (accessed June 21, 2011).

79 **the challenge came from an entrepreneur named Nick Merrill:** Kim Zetter,
"'John Doe' Who Fought FBI Spying Freed from Gag Order After 6 Years,"
Wired.com, August 10, 2010, www.wired.com/threatlevel/2010/08/nsl-gag
-order-lifted (accessed June 21, 2011).

79 **Christopher Soghoian, an antisurveillance activist and doctoral candidate at
Indiana University:** The paper cited is Christopher Soghoian, "The Law Enforcement Surveillance Reporting Gap," April 10, 2011, http://ssrn.com
/abstract=1806628.

79 **"intelligence investigations have compromised the civil liberties of American
citizens far more frequently, and to a greater extent, than was previously assumed":** Electronic Frontier Foundation, "Patterns of Misconduct: FBI Intelligence Violations from 2001–2008," January 2011, www.eff.org/pages
/patterns-misconduct-fbi-intelligence-violations.

80 **"Panopticon effect," named after a prison designed in 1785 by English
philosopher and social theorist Jeremy Bentham:** Jeremy Bentham, "Panopticon" (preface), in *The Panopticon Writings*, ed. Miran Bozovic (London:
Verso, 1995), 29–95.

80 **Michel Foucault warned that "panopticism" can extend beyond the physical
prison:** See Michel Foucault, *Discipline and Punish: The Birth of the Prison*
(New York: Vintage Books, 1995).

80 **In his recent book *One Nation, Under Surveillance*:** Simon Chesterman, *One
Nation, Under Surveillance: A New Social Contract to Defend Freedom Without
Sacrificing Liberty* (New York: Oxford University Press, 2011), 12.

81 **research paper comparing corporate data retention policies and companies'
specific practices in handing over user data:** Christopher Soghoian, "An End
to Privacy Theater: Exposing and Discouraging Corporate Disclosure of User
Data to the Government" (August 10, 2010), *Minnesota Journal of Law, Science
& Technology*, available at mjlst.umn.edu/previousissues/vol12iss1/home.html.

84 **Asked by the *New York Times* about the case, Twitter spokeswoman Jodi
Olson replied:** Scott Shane and John F. Burns, "US Subpoenas Twitter over
WikiLeaks Supporters," *New York Times*, January 8, 2011, www.nytimes
.com/2011/01/09/world/09wiki.html (accessed June 21, 2011).

85 **These included a letter by State Department legal adviser Harold Koh, in
which he wrote that the "violation of the law is ongoing":** "Text of State Department Letter to WikiLeaks," Reuters, November 28, 2010, www.reuters
.com/article/2010/11/28/us-wikileaks-usa-letter-idUSTRE6AR1E42010
1128 (accessed June 21, 2011).

85 **As Harvard legal scholar Yochai Benkler pointed out . . . Koh's assertion was
patently "false, as a matter of constitutional law":** This comment is from a
Berkman Center group e-mail exchange with Professor Benkler, quoted with
permission. For Benkler's in-depth analysis of the WikiLeaks case, see "A
Free Irresponsible Press: WikiLeaks and the Battle over the Soul of the

Networked Fourth Estate," *Harvard Civil Rights–Civil Liberties Law Review*, forthcoming; working draft http://benkler.org/Benkler_Wikileaks _current.pdf.

CHAPTER 6: DEMOCRATIC CENSORSHIP

87   **In a blog post explaining her decision:** Dan Frost, "The Attack on Kathy Sierra," *SFGate*, March 27, 2007, www.sfgate.com/cgi-bin/blogs/techchron /detail?entry_id=14783 (accessed June 27, 2011).

88   *The Offensive Internet*, **published in early 2011:** Danielle Keats Citron, "Civil Rights in our Information Age," and Cass R. Sunstein, "Believing False Rumors," in *The Offensive Internet: Speech, Privacy, and Reputation*, ed. Saul Levmore and Martha C. Nussbaum (Cambridge, MA: Harvard University Press, 2010), 31–49, 91–106.

89   **Constitutional lawyer Lee Bollinger:** His book is Lee C. Bollinger, *Uninhibited, Robust, and Wide-Open: A Free Press for a New Century (Inalienable Rights)* (New York: Oxford University Press, 2010), 48.

90   **"dog poop girl":** Jonathan Krim, "Subway Fracas Escalates into Test of the Internet's Power to Shame," *Washington Post*, July 7, 2005, www.washington post.com/wp-dyn/content/article/2005/07/06/AR2005070601953.html (accessed August 3, 2011).

90   **Cyber-harassment has already caused a number of celebrity suicides:** "Cyber Bullying Campaign Against Korean Singer Dies Down," Agence France-Presse, October 13, 2010, www.google.com/hostednews/afp/article/ALeqM5 ig4StQI4mbvccWeFCGC5uuihUyAg?docId=CNG.9dd1a1176881e712993 720a765eec626.1d1 (accessed August 3, 2011).

90   **Anonymity, South Korean legislators had come to believe, was undermining social stability:** See Byongil Oh, "Republic of Korea," *Global Information Society Watch 2009*, Association for Progressive Communications and Humanist Institute for Cooperation with Developing Countries, 150–152, www .apc.org/system/files/GISW2009Web_EN.pdf.

91   **In early 2009, South Korean blogger Park Dae-sung, aka Minerva, was arrested and jailed:** See Matthias Schwartz, "The Troubles of Korea's Influential Economic Pundit," *Wired Magazine*, October 19, 2009, www.wired .com/magazine/2009/10/mf_minerva (accessed August 3, 2011).

91   **seventeen people were charged with "spreading false information":** Joint Korean NGOs for the Official Visit of the Special Rapporteur to the Republic of Korea, NGO Report on the Situation of Freedom of Opinion and Expression in the Republic of Korea Since 2008, April 2010, http://kctu .org/7978.

91   **This law . . . prompted Google to disable uploading or comments on its Korean YouTube service in 2009:** Stephen Shankland, "YouTube Korea Squelches Uploads, Comments," CNET, April 13, 2009, http://news.cnet .com/8301-1023_3-10218419-93.html (accessed August 12, 2011).

91   the people of South Korea learned a painful lesson: See "Nate, Cyworld
     Hack Stole Information from 35 Million Users: SKorea Officials," Associated
     Press, July 28, 2011, www.huffingtonpost.com/2011/07/28/south-korea-nate
     -cyworld-hack-attack_n_911761.html; "Korean National ID Numbers
     Spring Up All over Chinese Web," *Korea Herald*/Asia News Network, August
     4, 2011, http://news.asiaone.com/News/Latest%2BNews/Science%2Band
     %2BTech/Story/A1Story20110804-292648.html; Kang Hyun-kyung,
     "Online Real Names to Be Scrapped," *Korea Times*, August 11, 2011, www
     .koreatimes.co.kr/www/news/nation/2011/08/116_92608.html (all accessed
     August 12, 2011).

92   A new law that went into effect in late 2009: An annotated copy of the full
     text of the Information Technology (Amendment) Act of 2008 can be found
     at http://cyberlaws.net/itamendments/IT%20ACT%20AMENDMENTS
     .PDF. Also see Amol Sharma and Jessica Vascellaro, "Google and India Test
     the Limits of Liberty," *Wall Street Journal*, January 4, 2010, http://online.wsj.com
     /article/SB126239086161213013.html (accessed August 3, 2011).

92   in April 2011, the Ministry of Communications and Information Technol-
     ogy went several steps further: Vikas Bajaj, "India Puts Tight Leash on In-
     ternet Free Speech," *New York Times*, April 27, 2011, www.nytimes.com
     /2011/04/28/technology/28internet.html (accessed August 3, 2011); and
     Pranesh Prakash, "Rebuttal of DIT's Misleading Statements on New In-
     ternet Rules," Centre for Internet and Society (Bangalore), May 13, 2011,
     http://cis-india.org/internet-governance/blog/rebuttal-dit-press-release
     -intermediaries.

93   Google publicly protested the rules: Rama Lakshmi, "India's New Internet
     Rules Criticized," *Washington Post*, August 1, 2011, http://www.washington
     post.com/world/indias-new-internet-rules-criticized/2011/07/27/gIQA1zS
     2mI_story.html (accessed August 3, 2011).

93   In early 2010 an Italian judge handed down criminal sentences to four top
     Google executives: Rachel Donadio, "Larger Threat Is Seen in Google
     Case," *New York Times*, February 24, 2010, www.nytimes.com/2010/02/25
     /technology/companies/25google.html (accessed August 3, 2011).

94   In a January 2011 report titled *The Slide from "Self Regulation" to Corporate
     Censorship*, the Brussels-based nonprofit European Digital Rights Ini-
     tiative (EDRI): The full report by Joe McNamee can be downloaded at
     www.edri.org/files/EDRI_selfreg_final_20110124.pdf.

95   the IWF's "procedures and policies are not transparent": Freedom House,
     Freedom on the Net: A Global Assessment of Internet and Digital Media, April 1,
     2009, 108, downloadable at www.freedomhouse.org/uploads/specialreports
     /NetFreedom2009/FreedomOnTheNet_FullReport.pdf. The 2011 report can
     be downloaded at www.freedomhouse.org/images/File/FotN/FOTN2011.pdf.

95   the 2008 book *Access Denied*: Ronald J. Deibert, et al., *Access Denied: The Prac-
     tice and Policy of Global Internet Filtering* (Cambridge, MA: MIT Press, 2008),
     186.

96   proposal to create a "single European cyberspace" that would block "illicit
      content" at Europe's borders: "Outcome of Proceedings: Joint Meeting of
      the Law Enforcement Working Party and the Customs Cooperation Work-
      ing Party on February 17, 2011," Council of the European Union, March 3,
      2011, 4, http://register.consilium.europa.eu/pdf/en/11/st07/st07181.en11
      .pdf. Also see Stewart Mitchell, "Dossier Reveals Proposals for EU-wide
      Content Blocking," *PC Pro*, May 12, 2011, www.pcpro.co.uk/news/367318
      /dossier-reveals-proposals-for-eu-wide-content-blocking (accessed June 27,
      2011).

96   in early 2011 Malcolm Hutty, president of the European Service Providers'
      Association, wrote a letter to the European Parliament calling for an end to
      Internet filtering, calling it an "inefficient measure": Chris Williams, "ISPs
      Battle EU Child Pornography Filter Laws," *The Register*, January 12, 2011,
      www.theregister.co.uk/2011/01/12/euroispa_eu/ (accessed June 27, 2011).

96   members of the European Parliament pointed out that a website campaign-
      ing *against* child pornography was blocked twice in the Netherlands: "Child
      Pornography: MEPs Doubt Effectiveness of Blocking Web Access," Euro-
      pean Parliament, November 15, 2010, www.europarl.europa.eu/en/pressroom
      /content/20101115IPR94729/html/Child-pornography-MEPs-doubt
      -effectiveness-of-blocking-web-access (accessed June 27, 2011).

96   research paper titled "Internet Blocking: Balancing Cybercrime Responses
      in Democratic Societies": Cormac Callanan, Marco Gercke, Estelle De
      Marco, and Hein Dries-Ziekenheiner, "Internet Blocking: Balancing Cy-
      bercrime Responses in Democratic Societies," Aconite Internet Solutions,
      October 2009, www.aconite.com/sites/default/files/Internet_blocking_and
      _Democracy.pdf.

96   WikiLeaks published a secret government list of 2,935 websites that Inter-
      net service providers would be required to block as part of a test run: Liam
      Thung, "WikiLeaks Spills ACMA Blacklist," *ZDNet*, March 19, 2009,
      www.zdnet.com.au/wikileaks-spills-acma-blacklist-339295538.htm (accessed
      June 27, 2011).

CHAPTER 7: COPYWARS

99   "The Google Predicament: Transforming US Cyberspace Policy to Advance
      Democracy, Security, and Trade": Full transcript of the congressional hear-
      ing can be downloaded at www.fas.org/irp/congress/2010_hr/google.pdf.

99   Google Vice President and Deputy General Counsel Nicole Wong. Two
      years earlier, the *New York Times* had nicknamed her "the decider": Jeffrey
      Rosen, "Google's Gatekeepers," *New York Times Magazine*, November 28,
      2008, www.nytimes.com/2008/11/30/magazine/30google-t.html (accessed
      June 27, 2011).

101   In a January 2010 op-ed in the *New York Times*: Bono, "Ten for the Next Ten,"
      New York Times, January 2, 2010, www.nytimes.com/2010/01/03/opinion
      /03bono.html (accessed June 27, 2011).

101 exactly what Foreign Ministry spokesperson Jiang Yu said: "Chinese Government's Management of Internet Is Sovereign Act, Says FM," Xinhua News Agency, May 19, 2011, http://news.xinhuanet.com/english2010 /china/2011-05/19/c_13883854.htm (accessed June 27, 2011).

102 Preventing Real Online Threats to Economic Creativity and Theft of Intellectual Property Act of 2011, otherwise known by its acronym, PROTECT IP: For the legislation's full text and status, see www.govtrack.us/congress /bill.xpd?bill=s112-968.

102 degrade "the Internet's value as a single, unified, global communications network": Steve Crocker, David Dagon, Dan Kaminsky, Danny McPherson, and Paul Vixie, "Security and Other Technical Concerns Raised by the DNS Filtering Requirements in the PROTECT IP Bill," May 2011, www.redbarn .org/files_redbarn/PROTECT-IP-Technical-Whitepaper-Final.pdf.

102 Civil liberties and free speech groups argued that the bill lacked safeguards: See, for example, Declan McCullagh, "Senate Bill Amounts to Death Penalty for Web Sites," CNet News, May 12, 2011, http://news.cnet.com/8301 -31921_3-20062398-281.html; and David Sohn, "Copyright Bill Advances, but Draws Plenty of Criticism," Center for Democracy and Technology blog, May 26, 2011, www.cdt.org/blogs/david-sohn/copyright-bill-advances-draws -plenty-criticism.

102 "If protecting intellectual property is important, so is protecting the Internet from overzealous enforcement": "Internet Piracy and How to Stop It," New York Times, June 8, 2011, www.nytimes.com/2011/06/09/opinion /09thu1.html (accessed August 8, 2011).

102 In November 2010 the US Department of Homeland Security's Immigration and Customs Enforcement unit (known as ICE) shut down eighty-two websites: Ben Sisario, "Music Web Sites Dispute Legality of Their Closing," New York Times, December 19, 2010, www.nytimes.com/2010/12/20 /business/media/20music.html (accessed June 27, 2011).

103 "Takedown Hall of Shame": See www.eff.org/takedowns.

104 Ars Technica found its popular Facebook page shut down in May 2011 as the result of a DMCA complaint: Ken Fisher, "Facebook Shoots First, Ignores Questions Later; Account Lock-Out Attack Works (Update X)," Ars Technica, April 28, 2011, http://arstechnica.com/business/news/2011/04/facebook -shoots-first-ignores-questions-later-account-lock-out-attack-works.ars; and Jacqui Cheng, "Facebook Takedown Followup: What Happened, and What Facebook Needs to Fix," Ars Technica, April 29, 2011, http://arstechnica .com/tech-policy/news/2011/04/facebook-takedown-followup-what -happened-and-what-facebook-needs-to-fix.ars (both accessed June 27, 2011).

105 Chinese authorities routinely combine intellectual property enforcement campaigns with broader efforts to stamp out not only pornography but also antigovernment material: Andrew Mertha, The Politics of Piracy: Intellectual Property in Contemporary China (New York: Cornell University Press, 2005), 134–144.

105 authorities viewed American pressure to tighten intellectual property enforcement and improve copyright law as politically convenient: Stephen McIntyre, "The Yang Obeys, but the Yin Ignores: Copyright Law and Speech Suppression in the People's Republic of China," *UCLA Pacific Basin Law Journal*, 2011, http://ssrn.com/abstract=1752443.

106 US trade policies pushing for tougher enforcement have been blind: For more specifics see Tome Broud, "It's Easily Done: The China-Intellectual Property Rights Enforcement Dispute and the Freedom of Expression," *World Journal of Intellectual Property* 13, no. 5 (September 2010): 660–673.

106 ACTA's Internet section: For links to various leaked drafts as well as the text as officially released in April 2010, see Gwen Hinze, "Preliminary Analysis of the Officially Released ACTA Text," EFF Deeplinks Blog, April 22, 2010, www.eff.org/deeplinks/2010/04/eff-analysis-officially-released-acta-text (accessed June 27, 2011).

106 WikiLeaks obtained a leaked copy and published it online: See Grant Gross, "EFF, Public Knowledge Sue US Gov't over Secret IP Pact," *InfoWorld*, September 18, 2008, www.infoworld.com/d/security-central/eff-public-knowledge -sue-us-govt-over-secret-ip-pact-955 (accessed June 27, 2011).

107 report by the Social Science Research Council examining the impact of US intellectual property enforcement trade policies: Joe Karaganis, ed., "Media Piracy in Emerging Economies," Social Science Research Council, March 2011, http://piracy.ssrc.org.

107 In September 2010, the *New York Times* broke the story that Russian security services had carried out dozens of raids: See Clifford J. Levy, "Russia Uses Microsoft to Suppress Dissent," *New York Times*, September 11, 2010, www.nytimes.com/2010/09/12/world/europe/12raids.html; and Clifford J. Levy, "Undercut by Microsoft, Russia Drops Piracy Case," December 5, 2010, www.nytimes.com/2010/12/06/world/europe/06russia.html (both accessed June 27, 2011).

108 Sarkozy's government has created a new agency, HADOPI: See Eric Pfanner, "France Approves Wide Crackdown on Net Piracy," *New York Times*, October 22, 2009, www.nytimes.com/2009/10/23/technology/23net.html. HADOPI's proponents cite a recent Ministry of Culture survey that points to a change in public behavior as a result of the new measures. See Aymeric Pichevin, "HADOPI Study Says France's Three-Strike Law Having Positive Impact on Music Piracy," Billboard.biz, May 16, 2011, www.billboard.biz /bbbiz/industry/digital-and-mobile/hadopi-study-says-france-s-three-strike -1005185142.story (both accessed June 27, 2011).

108 the company, which was contracted to serve as a clearinghouse for user information that other companies collected to track infringement, was hacked: Nate Anderson, "France Halts 'Three Strikes' IP Address Collection After Data Leak," *Ars Technica*, May 17, 2011, http://arstechnica.com/tech-policy /news/2011/05/france-halts-three-strikes-ip-address-collection-after-data -leak.ars (accessed June 27, 2011).

108   **Broadband analyst Mark Jackson warned:** Matthew Richardson, "Illegal File Sharing Cannot Be Stopped, Says Broadband Commentator," SimplifyDigital, December 21, 2010, www.simplifydigital.co.uk/news/articles/2010/12 /illegal-file-sharing-cannot-be-stopped-says-broadband-commentator (accessed June 27, 2011).

109   **"lobbynomics":** Ian Hargreaves, "Digital Opportunity: A Review of Intellectual Property and Growth," May 2011, www.ipo.gov.uk/ipreview.htm.

109   **the recorded music industry, led by the Recording Industry Association of America, spent a combined $17.5 million in congressional lobbying in 2009 alone:** Bruce Gain, "Special Report: Music Industry's Lavish Lobby Campaign for Digital Rights," Intellectual Property Watch, January 6, 2011, www.ip-watch.org/weblog/2011/01/06/special-report-music-industrys -lavish-lobby-campaign-for-digital-rights (accessed June 27, 2011).

110   **policy outcomes strongly reflect the preferences of the most affluent:** Martin Gilens, "Inequality and Democratic Responsiveness," *Public Opinion Quarterly* 69, no. 5 (Special Issue 2005): 778–796, http://poq.oxfordjournals.org /content/69/5/778.full.pdf.

110   **Until this fundamental flaw in American democracy is addressed:** To that end, Lessig is leading two new, related projects: Fix Congress First (www .fixcongressfirst.org) and Root Strikers (www.rootstrikers.org).

111   **Tim Berners-Lee . . . suggested that it is time for humanity to upgrade the principles first articulated eight hundred years ago in the Magna Carta:** Tim Berners-Lee, "Long Live the Web: A Call for Continued Open Standards and Neutrality," *Scientific American*, November 22, 2010, www.scientific american.com/article.cfm?id=long-live-the-web.

CHAPTER 8: CORPORATE CENSORSHIP

115   **On Apple's special store for the Chinese market, apps related to the Dalai Lama are censored:** Owen Fletcher, "Apple Censors Dalai Lama iPhone Apps in China," *IDG News*, December 29, 2009, www.pcworld.com/article /185604/apple_censors_dalai_lama_iphone_apps_in_china.html (accessed June 27, 2011).

116   **When challenged, company executives said:** See Anita Ramasastry, "Should Cellphone Companies Be Able to Censor the Messages We Send? The Verizon/NARAL Controversy," *FindLaw*, October 11, 2007, http://writ .lp.findlaw.com/ramasastry/20071011.html; and Kim Hart, "Verizon Ends Text-Message Ban," *Washington Post*, September 28, 2007, www.washington post.com/wp-dyn/content/article/2007/09/27/AR2007092700823.html (both accessed June 27, 2011).

117   **Though the group sued T-Mobile, the company called the lawsuit "without merit":** See Verne G. Kopytoff, "T-Mobile Sued over Blockade of Text Messages," *New York Times* Bits Blog, September 20, 2010, http://bits.blogs .nytimes.com/2010/09/20/t-mobile-blocks-text-messages (accessed June 27, 2011).

117   during a live-streaming broadcast of a Pearl Jam concert, AT&T muted the
       sound: See Staci D. Kramer, "AT&T Silences Pearl Jam; Gives 'Net Neutral-
       ity' Proponents Ammunition," Forbes.com, August 9, 2007, www.forbes
       .com/2007/08/09/att-pearljam-music-tech-cx_pco_0809paidcontent.html
       (accessed June 27, 2011).

117   Comcast has been accused of blocking entire peer-to-peer applications: See
       K. C. Jones, "FCC Orders Comcast to Stop Blocking Internet Traffic," In-
       formationWeek, August 1, 2008, www.informationweek.com/news/internet
       /policy/209901533.

118   Level 3 accused Comcast: Amy Tomson and Todd Shields, "Comcast Starts
       Web 'Toll Booth,' Netflix Supplier Says," Bloomberg, November 30, 2010,
       www.bloomberg.com/news/2010-11-30/comcast-starts-online-video-toll
       -booth-netflix-web-partner-level-3-says.html (accessed June 27, 2011).

118   On a "neutral" Internet: For a thorough analysis of the free speech issues re-
       lated to net neutrality, see Dawn C. Nunziato, Virtual Freedom: Net Neutral-
       ity and Free Speech in the Internet Age (Palo Alto, CA: Stanford Law Books,
       2009).

119   The Master Switch: Tim Wu, The Master Switch: The Rise and Fall of Infor-
       mation Empires (New York: Random House, 2010).

119   prerequisite for Internet freedom: See Al Franken, "Net Neutrality Is Foremost
       Free Speech Issue of Our Time," CNN.com, August 5, 2010, http://articles
       .cnn.com/2010-08-05/opinion/franken.net.neutrality_1_net-neutrality
       -television-networks-cable.

120   Opponents of net neutrality: For one of many critiques of net neutrality, see
       Adam D. Theirer, "'Net Neutrality': Digital Discrimination or Regulatory
       Gamesmanship in Cyberspace?" Cato Institute, January 12, 2004, www
       .cato.org/pubs/pas/pa-507es.html. Also see Robert M. McDowell, "The FCC's
       Threat to Internet Freedom," Wall Street Journal, December 19, 2010, http://
       online.wsj.com/article/SB10001424052748703395204576023452250748540
       .html

120   As The Economist pointed out in December 2010: "Network Neutrality: A
       Tangled Web," The Economist, December 29, 2010, www.economist.com
       /node/17800141 (accessed June 27, 2011).

121   in 2010 Chile became the first nation to enshrine net neutrality into law:
       Iain Thomson, "Chile Becomes First Net Neutrality Nation," July 16, 2010,
       www.v3.co.uk/v3-uk/news/1961015/chile-net-neutrality-nation (accessed
       June 27, 2011).

121   In late 2010 the European Parliament and European Commission held a
       joint net neutrality summit: See "Results of Public Consultation on Open
       Internet and Net Neutrality," European Commission, November 2010, http://
       ec.europa.eu/information_society/policy/ecomm/library/public_consult/net
       _neutrality/index_en.htm.

121   as companies grow more comfortable with overtly interfering with customer
       communications: See Joe McNamee, "Net Neutrality—Wait and See the

End of the Open Internet," *EDRi-gram* no. 8.22, November 17, 2010, www
.edri.org/book/export/html/2444.

121 **the Netherlands became the first European country to make net neutrality
the law:** Kevin O'Brien, "Dutch Lawmakers Adopt Net Neutrality Law,"
New York Times, June 22, 2011, www.nytimes.com/2011/06/23/technology
/23neutral.html (accessed June 27, 2011).

122 **ISP had granted itself "powers that correspond to the Peruvian administrative and judicial authorities":** Jorge Bossio Montes de Oca, "Peru: The Battle for Control of the Internet," Association for Progressive Communications,
June 2009, www.apc.org/en/system/files/CILACInvestigacionesPeru_EN
_20090630.pdf.

122 **Kenyan engineer and blogger Tom Makau:** Tom Makau, "Network Neutrality: The African Perspective," Tom Makau: My Take on the Telecommunications and ICT industry, November 19, 2010, http://tommakau.com
/2010/11/19/network-neutrality-the-african-perspective.

122 **Google and Verizon had worked out a common position on net neutrality
policy:** Alan Davidson and Tom Tauke, "A Joint Policy Proposal for an Open
Internet," Google Public Policy Blog, August 9, 2010, http://googlepublic
policy.blogspot.com/2010/08/joint-policy-proposal-for-open-internet.html;
and Brian Stelter, "FCC Is Set to Regulate Net Access," *New York Times*,
December 20, 2010, www.nytimes.com/2010/12/21/business/media/21fcc
.html (accessed June 27, 2011).

123 **Facebook Zero:** See "Facebook Launch 'Zero' Site for Mobile Phones," BBC
News, February 16, 2010, http://news.bbc.co.uk/2/hi/technology/8518
726.stm. For Facebook's official description, see Sid Murlidhar, "Fast and
Free Facebook Mobile Access with 0.facebook.com," Facebook Blog, May
18, 2010 (accessed June 27, 2011).

125 **MetroPCS:** See Ryan Singel, "MetroPCS 4G Data-Blocking Plans May Violate Net Neutrality," Wired.com, January 7, 2011, www.wired.com/epicenter
/2011/01/metropcs-net-neutrality; and Ryan Singel, "Court Tosses Net Neutrality Challenges—For Now," April 4, 2011, www.wired.com/epicenter
/2011/04/net-neutrality-challenges-tossed (both accessed June 27, 2011).

126 **Apple shut down, without notice, an iPad application for *Stern*:** See Eric
Pfanner, "Publishers Question Apple's Rejection of Nudity," *New York Times*,
March 14, 2010, www.nytimes.com/2010/03/15/technology/15cache.html
(accessed June 27, 2011).

126 **Apple also censors controversial political and religious content:** See, for example, Ryan Singel, "New Yorker's Remnick Says He Won't Censor to Make
Apple Happy," Wired.com, May 25, 2010, www.wired.com/epicenter
/2010/05/new-yorker-apple; and Matthew Shaer, "Guy Wins Pulitzer, Finally Has iPhone App Accepted by Apple," *Christian Science Monitor*, April
16, 2010, www.csmonitor.com/Innovation/Horizons/2010/0416/Guy
-wins-Pulitzer-finally-has-iPhone-app-accepted-by-Apple (both accessed
June 27, 2011).

127   **Apple . . . pulled an app by the Christian group Exodus International:** See
        David Crary, "Some Gay-Rights Foes Claim They Now Are Bullied," Asso-
        ciated Press, June 11, 2011; and Ki Mae Heussner, "Apple Pulls 'Anti-Gay'
        App After Pressure," ABCNews.com, March 23, 2011, http://abcnews
        .go.com/Technology/apple-pulls-anti-gay-app-thousands-sign-petition
        /story?id=13201555&page=2 (accessed June 27, 2011).

129   **the app was also collecting the user's personal data and sending it to a "mys-
        terious site in China":** Dean Takahashi, "Updated: Android Wallpaper App
        That Takes Your Data Was Downloaded by Millions," VentureBeat, July 28,
        2010, http://venturebeat.com/2010/07/28/android-wallpaper-app-that-steals
        -your-data-was-downloaded-by-millions. Also see Brian Prince, "Google
        Android Malware Threat Targets App Market," e-Week.com, March 2, 2011,
        www.eweek.com/c/a/Security/Google-Android-Malware-Threat-Targets
        -App-Market-466899 (both accessed June 27, 2011).

129   **Apple . . . calling Android apps "inferior":** Gregg Keizer, "Apple Slams Ama-
        zon's Android E-store as 'Inferior,'" June 10, 2011, www.computerworld
        .com/s/article/9217533/Apple_slams_Amazon_s_Android_e_store_as
        _inferior. Also see Byron Acohido, "Android, Apple Face Growing Cyberat-
        tacks," *USA Today,* June 6, 2011, www.usatoday.com/tech/products/2011-06
        -03-tougher-security-sought-in-google-apple-devices_n.htm (accessed June
        27, 2011).

CHAPTER 9: DO NO EVIL

131   **In April 2011, Mike Lazaridis, co-CEO of Research in Motion (RIM),
        maker of BlackBerry, sat down for an interview with the BBC's Rory
        Cellan-Jones:** Full video of the exchange is at http://news.bbc.co.uk/2/hi
        /programmes/click_online/9456798.stm (accessed June 27, 2011).

133   **Shi Tao:** For a detailed account and list of sources related to the Shi Tao case
        and Yahoo, see Human Rights Watch, *Race to the Bottom: Corporate Complic-
        ity in Chinese Internet Censorship,* 2006, www.hrw.org/reports/2006/china
        0806; and Rebecca MacKinnon, "Shi Tao, Yahoo!, and the Lessons for Cor-
        porate Social Responsibility," working paper, December 27, 2007, http://
        rconversation.blogs.com/YahooShiTaoLessons.pdf.

136   **the year Microsoft launched MSN Spaces in China, 2005, was also the year
        the Chinese blogosphere exploded:** For a detailed account of the evolution of
        the Chinese blogosphere and government controls, see Rebecca MacKinnon,
        "Flatter World and Thicker Walls? Blogs, Censorship, and Civic Discourse
        in China," in D. Drezner and H. Farrell, eds., Special Issue: Will the Revo-
        lution Be Blogged? *Public Choice* 134, no. 1–2 (January 2008): 31–46.

140   **Ebele Okobi-Harris, director of Yahoo's business and human rights pro-
        gram, explained in a blog post:** Ebele Okobi-Harris, "Thoughts on Flickr
        and Human Rights," Yahoo Business and Human Rights Program Blog,
        March 15, 2011, www.yhumanrightsblog.com/blog/2011/03/15/thoughts
        -on-flickr-and-human-rights (accessed June 27, 2011).

140   Veteran activist Gilles Frydman, who had brought the issue to public atten-
      tion a few days previously, challenged Okobi-Harris outright at the confer-
      ence: I was present at the conference panel during which this exchange took
      place. Also see Jennifer Preston, "Ethical Quandary for Social Sites," *New
      York Times*, March 27, 2011, www.nytimes.com/2011/03/28/business/media
      /28social.html (accessed June 27, 2011).

141   it is more like a shopping mall or a food court: "These entities have no more
      legal obligation to allow open, unfettered political speech in their spaces than
      shopping malls do to host political rallies." Ethan Zuckerman, "Internet
      Freedom: Protect, Then Project," My Heart's in Accra blog, March 22, 2010,
      www.ethanzuckerman.com/blog/2010/03/22/internet-freedom-protect
      -then-project.

142   "My privacy concerns are not trite": Harriet J, "Fuck You, Google," Fugitivus,
      February 11, 2010, www.fugitivus.net/2010/02/11/fuck-you-google.

145   Iranian users were deleting their accounts in horror: Comment posted by
      "marblehead160" at www.zdnet.com/tb/1-72584-1405690 in response to
      Zack Whittaker, "Facebook Will Never Get Privacy Right," ZDNet, De-
      cember 10, 2009, www.zdnet.com/blog/igeneration/facebook-will-never-get
      -privacy-right/3588 (both accessed June 27, 2011).

146   As privacy advocate Kaliya Hamlin put it: Kaliya Hamlin, "Facebook's Pri-
      vacy Move Violates Contract with Users," ReadWriteWeb, December 15,
      2009, www.readwriteweb.com/archives/facebooks_privacy_move_violates
      _contract_with_user.php (accessed June 27, 2011).

CHAPTER 10: FACEBOOKISTAN AND GOOGLEDOM

149   Lokman Tsui . . . likened Facebook to a country run by an authoritarian,
      paternalistic government: Lokman Tsui, "Dear Facebook, Freedom or
      Friends? That's Not a Choice," Global Voices, One World, May 6, 2010,
      www.lokman.org/2010/05/06/dear-facebook-freedom-or-friends-thats-not
      -a-choice.

150   "radical transparency": David Kirkpatrick, *The Facebook Effect: The Inside
      Story of the Company That Is Connecting the World* (New York: Simon and
      Schuster, 2011), 209–210, 199.

151   in June 2009 a young man named Khaled Said: See Ramy Raoof, "Egypt:
      Facebook Disables Popular Anti-Torture Page," Global Voices Advocacy, No-
      vember 25, 2010, http://advocacy.globalvoicesonline.org/2010/11/25/egypt
      -facebook-disables-popular-anti-torture-page; and Jennifer Preston, "Move-
      ment Began with Outrage and a Facebook Page That Gave It an Outlet,"
      New York Times, February 5, 2011, www.nytimes.com/2011/02/06/world/
      middleeast/06face.html (accessed June 27, 2011).

154   The Simon Wiesenthal Center is unhappy: See Miguel Helft, "Facebook
      Wrestles with Free Speech and Civility," *New York Times*, December 12,
      2010, www.nytimes.com/2010/12/13/technology/13facebook.html (accessed
      June 27, 2011).

154   Zuckerberg alluded to a kind of social contract between Facebook and the user: Kirkpatrick, *The Facebook Effect*, 100.

155   "lack of control that exists on the Internet as a whole": Tim Bradshaw and Joseph Menn, "Facebook to Simplify Its Controls," *Financial Times*, May 14, 2010, www.ft.com/intl/cms/s/0/d93102f4-5f81-11df-a670-00144feab49a .html (accessed June 27, 2011).

155   a Facebook page with more than 800,000 members called "Boycott BP" . . . was disabled: "Facebook Says It Disabled 'Boycott BP' Page in Error," CNN .com, June 29, 2010, http://articles.cnn.com/2010-06-29/tech/facebook .bp.controversy_1_oil-giant-bp-bp-spokesman-facebook (accessed June 27, 2011).

157   "high time that we take into consideration those whose lives aren't nearly as privileged as ours": danah boyd, "Facebook and 'Radical Transparency' (a Rant)," danah boyd | apophenia, May 14, 2010, www.zephoria.org/thoughts /archives/2010/05/14/facebook-and-radical-transparency-a-rant.html.

157   "We will soon ramp up our efforts to provide better guidance to those confused": "Facebook Executive Answers Reader Questions," *New York Times* Bits Blog, May 11, 2010, http://bits.blogs.nytimes.com/2010/05/11 /facebook-executive-answers-reader-questions.

158   danah boyd wrote an essay arguing that activism rather than boycott is likely to be more effective: danah boyd, "Quitting Facebook Is Pointless; Challenging Them to Do Better Is Not," danah boyd | apophenia, May 23, 2010, www.zephoria.org/thoughts/archives/2010/05/23/quitting-facebook -is-pointless-challenging-them-to-do-better-is-not.html.

159   Facebook's engineers added new encryption and security settings: Danny O'Brien, "Facebook Enables Encryption: A First Step on the Right Road," Committee to Protect Journalists, January 26, 2011, www.cpj.org/internet /2011/01/facebook-turns-on-https-a-first-step-on-the-right.php; and Juan Carlos Perez, "Facebook Tightens Log-in Verification," IDG News, May 12, 2011, www.pcworld.com/businesscenter/article/227786/facebook_tightens _login_verification.html (accessed August 11, 2011). Also, for an activists' guide to using Facebook's privacy and security features, see Susannah Vila, "How to Organize on Facebook Securely," www.movements.org/how-to /entry/organize-on-facebook-securely.

159   in mid-2011 the company rolled out an easy-to-use appeals process: Jillian York, "Facebook Appeals: We Has Them!" Jillian C. York blog, July 19, 2011, http://jilliancyork.com/2011/07/19/facebook-appeals.

160   Many welcomed Google Plus's more sophisticated approach to privacy: Ryan Singel, "Google Plus vs. Facebook on Privacy: Plus Ahead on Points—For Now," Wired.com, June 29, 2011, www.wired.co.uk/news/archive/2011-06/29 /google-facebook-privacy; and Jillian York, "Community Standards: A Comparison of Facebook vs. Google+," Jillian C. York blog, June 30, 2011, http://jilliancyork.com/2011/06/30/google-vs-facebook (accessed August 11, 2011).

160   hoped that more competition would force all companies to improve their
        practices: Danny O'Brien, "Security vs. risk: More on Facebook and Google+,"
        Committee to Protect Journalists, July 12, 2011, www.cpj.org/internet
        /2011/07/security-vs-risk-more-on-facebook-and-google.php.

161   In mid-July Google moved to deactivate pseudonymous accounts en masse,
        without warning: Violet Blue, "Google Plus Deleting Accounts En Masse:
        No Clear Answers," ZDNet, July 23, 2011, www.zdnet.com/blog/violetblue
        /google-plus-deleting-accounts-en-masse-no-clear-answers/567 (accessed
        August 11, 2011).

161   "GrrlScientist" wrote an article about her experience: GrrlScientist, "Google's
        Gormless 'No Pseudonym' Policy," Punctuated Equilibrium blog, *The Guardian*,
        July 25, 2011, www.guardian.co.uk/science/punctuated-equilibrium/2011
        /jul/25/1.

161   Skud had anticipated this situation: Skud, "An Update on My Google Plus
        Suspension," Infotropism blog, July 31, 2011, http://infotrope.net/2011
        /07/31/an-update-on-my-google-plus-suspension.

162   Randi Zuckerberg: Emil Protalinski, Facebook: "Anonymity on the Internet
        Has to Go Away," ZDNet, August 2, 2011, www.zdnet.com/blog/facebook
        /facebook-8220anonymity-on-the-internet-has-to-go-away-8221/2270 (ac-
        cessed August 11, 2011).

163   In early August, the company reaffirmed its real-ID policy: Marshall Kirk-
        patrick, "Google Plus Tells Pseudonym Lovers to Shove It," ReadWriteWeb,
        August 11, 2011, www.readwriteweb.com/archives/google_plus_tells
        _pseudonym_lovers_to_shove_it.php (accessed August 12, 2011).

CHAPTER 11: TRUST, BUT VERIFY

169   International Engineering Task Force: www.ietf.org.

169   Senior Vice President and General Counsel Mark Chandler . . . "our prod-
        ucts are built on open, global standards": Mark Chandler, "Cisco Supports Free-
        dom of Expression, an Open Internet and Human Rights," Cisco Blog, June 6,
        2011, http://blogs.cisco.com/news/cisco-supports-freedom-of-expression
        -an-open-internet-and-human-rights.

170   Cisco executives argue that their role in China and elsewhere has been
        misunderstood: Mark Chandler, "Cisco Testimony Before House Interna-
        tional Relations Subcommittee," Cisco Blog, February 16, 2006, http://
        blogs.cisco.com/gov/cisco_testimony_before_house_international_relations
        _subcommittee.

170   *Wall Street Journal* reported that Cisco . . . would supply some of the net-
        working equipment: Loretta Chao and Don Clark, "Cisco Poised to Help
        China Keep an Eye on Its Citizens," *Wall Street Journal*, July 5, 2011, http://
        online.wsj.com/article/SB10001424052702304778304576377141077267316
        .html (accessed August 13, 2011).

170   In a blog post responding to the article: Mark Chandler, "Cisco Responds to
        Wall Street Journal Article on China," Cisco blog, July 6, 2011, http://

blogs.cisco.com/news/cisco-responds-to-wall-street-journal-article
-on-china.

170   **In 2005 the author Ethan Gutmann published Chinese-language marketing brochures for surveillance equipment:** Gutmann wrote about how he obtained them in *Losing the New China: A Story of American Commerce, Desire, and Betrayal* (New York: Encounter Books, 2004), 167–170. For images of the brochures themselves, see Rebecca MacKinnon, "More on Cisco in China," RConversation blog, June 30, 2005, http://rconversation.blogs.com /rconversation/2005/06/more_on_cisco_i.html.

170   **In 2008 activists published a PowerPoint presentation by a Cisco marketing manager:** Sarah Lai Stirland, "Cisco Leak: 'Great Firewall' of China Was a Chance to Sell More Routers," Wired.com, www.wired.com/threatlevel /2008/05/leaked-cisco-do (accessed June 27, 2011).

171   **Boston Common Asset Management divested:** "Weak Commitment to Human Rights Factors into Boston Common's Decision to Divest of Cisco Systems; Manipulative Vote Tallying Further Isolates Cisco," press release issued by Boston Common Asset Management, January 10, 2011, http:// bostoncommonasset.com/news/documents/CiscoDivestmentStatement 011011.pdf.

174   **Global Online Freedom Act:** Text and status updates for HR 1389—Global Online Freedom Act of 2011 can be found at www.govtrack.us/congress /bill.xpd?bill=h112-1389.

176   **assets held by US investors in some form of socially responsible investment funds:** See "2010 Report on Socially Responsible Investing Trends in the United States," Social Investment Forum Foundation, http://ussif.org /resources/pubs; and "Socially Responsible Investing Facts," US-SIF: The Forum for Sustainable and Responsible Investment, http://ussif.org/resources /sriguide/srifacts.cfm.

176   **what Harvard Business School gurus Michael Porter and Mark Kramer call "shared value":** Michael E. Porter and Mark R. Kramer, "The Big Idea: Creating Shared Value," *Harvard Business Review*, January–February 2011, 62–77.

180   **Global Network Initiative:** http://globalnetworkinitiative.org.

181   **Yahoo's handling of its Vietnamese business:** See Colin M. Maclay, "Protecting Privacy and Expression Online: Can the Global Network Initiative Embrace the Character of the Net?" in *Access Controlled*, ed. Ronald J. Deibert et al. (Cambridge, MA: MIT Press, 2010), 87–88.

182   **Nokia Siemens, Ericsson, and Vodafone have all come under fire for assisting government suppression in Iran, Belarus, and Egypt, respectively:** See Christopher Rhoads and Loretta Chao, "Iran's Web Spying Aided by Western Technology," *Wall Street Journal*, June 22, 2009, http://online.wsj.com /article/SB124562668777335653.html; Hans Rosen, "Ericsson Technology Used to Wiretap in Belarus," *Dagens Nyheter*, December 22, 2010, www.dn .se/nyheter/varlden/ericsson-technology-used-to-wiretap-in-belarus; and

Raphael G. Satter, "Vodafone: Egypt Forced Us to Send Text Messages," Associated Press, February 3, 2011, www.businessweek.com/ap/financialnews /D9L5ANI80.htm (all accessed August 13, 2011).

185   **in June 2011 the UN Human Rights Council approved the Guiding Principles on Business and Human Rights:** "Report of the Special Representative of the Secretary-General on the Issue of Human Rights and Transnational Corporations and Other Business Enterprises, John Ruggie; Guiding Principles on Business and Human Rights: Implementing the United Nations 'Protect, Respect and Remedy' Framework," UN Human Rights Council A/HRC/17/31, March 21, 2011, www.business-humanrights.org/media /documents/ruggie/ruggie-guiding-principles-21-mar-2011.pdf.

CHAPTER 12: IN SEARCH OF "INTERNET FREEDOM" POLICY

189   Global Internet Freedom Consortium (GIFC): www.internetfreedom.org.

189   **The GIFC found powerful allies in Mark Palmer . . . and Michael Horowitz:** See John Markoff, "Iranians and Others Outwit Net Censors," *New York Times*, April 30, 2009, www.nytimes.com/2009/05/01/technology /01filter.html; James O'Toole, "Internet Censorship Fight Goes Global," *Pittsburgh Post-Gazette*, June 4, 2009, www.post-gazette.com/pg/09155 /974993-82.stm; Brad Stone, "Aid Urged for Groups Fighting Internet Censors," *New York Times*, January 20, 2010, www.nytimes.com/2010/01 /21/technology/21censor.html; Caylan Ford, "What Hillary Clinton, Google Can Do About Censorship in China," *Washington Post*, January 20, 2010, www.washingtonpost.com/wp-dyn/content/article/2010/01/20/AR2010 012002805.html; Genevieve Long, "Internet Freedom Software Should Get Federal Funding, Group Says," *Epoch Times*, March 5, 2010; Gordon Crovitz, "Mrs. Clinton, Tear Down This Cyberwall," *Wall Street Journal*, May 3, 2010, http://online.wsj.com/article/SB10001424052748704608104575219022492 475364.html; Laura Smith-Spark, "US 'To Give $1.5m to Falun Gong Internet Freedom Group,'" BBC News, May 12, 2010, http://news.bbc.co.uk/2 /hi/8678760.stm; and Vince Beiser, "Digital Weapons Help Dissidents Punch Holes in China's Great Firewall," *Wired*, November 1, 2010, www .wired.com/magazine/2010/11/ff_firewallfighters.

190   **in January 2011 [the State Department] announced:** E. B. Boyd, "The State Department Has $30 Million to Spend on Internet Freedom," *Fast Company*, January 5, 2011, www.fastcompany.com/1714260/the-state-department-has -30-million-to-spend-on-internet-freedom. For the State Department's Request for Statements of Interest, see Craig Zelizer, "US Department of State, Call for Expressions of Interest, Internet Freedom Programs," Peace and Collaborative Development Network, January 3, 2011, www.international peaceandconflict.org/forum/topics/us-department-of-state-call-1.

190   **a much broader range of threats:** In my own writings and congressional testimony, I argued that circumvention solved only one of the many threats

online activists face, and I supported a much broader funding strategy. See http://judiciary.senate.gov/pdf/10-03-02MacKinnon%27sTestimony.pdf and http://rconversation.blogs.com/MacKinnonHFAC_March10.pdf. Also see Rebecca MacKinnon, "No Quick Fixes for Internet Freedom," *Wall Street Journal Asia*, November 19, 2010, http://online.wsj.com/article/SB10001 424052748704104104575622080860055498.html; and "Q. & A. with Rebecca MacKinnon: Internet in China" with Evan Osnos, *New Yorker*, February 22, 2011, www.newyorker.com/online/blogs/evanosnos/2011/02/internet -in-china.html.

190 **Senator Richard Lugar . . . called for the remaining funds to be removed from State Department control and given to the Broadcasting Board of Governors (BBG):** "Another U.S. Deficit—China and America—Public Diplomacy in the Age of the Internet: A Minority Staff Report Prepared for the Use of the Committee on Foreign Relations, United States Senate," February 15, 2011, http://lugar.senate.gov/issues/foreign/diplomacy/China Internet.pdf.

191 **"Internet-in-a-suitcase":** James Glanz and John Markoff, "US Underwrites Internet Detour Around Censors," *New York Times*, June 12, 2011, www .nytimes.com/2011/06/12/world/12internet.html. Also see Josh Smith, "State Allocates Final $28 Million for Internet Freedom Programs," *National Journal*, May 3, 2011, www.nextgov.com/nextgov/ng_20110503_8059.php.

191 **Clay Shirky critiqued Washington's obsession with circumvention:** Clay Shirky, "The Political Power of Social Media," *Foreign Affairs* 90, no. 1 (January–February 2011): 28–41.

193 **as Ethan Zuckerman of Harvard's Berkman Center warns:** See Ethan Zuckerman, "Internet Freedom: Beyond Circumvention," My Heart's in Accra blog, February 22, 2010, www.ethanzuckerman.com/blog/2010/02/22/internet -freedom-beyond-circumvention.

193 **Evgeny Morozov has been even more critical:** Evgeny Morozov, *The Net Delusion: The Dark Side of Internet Freedom* (New York: PublicAffairs, 2011).

193 **US Internet freedom policy also has critics among its intended beneficiaries:** Sami Ben Gharbia, "The Internet Freedom Fallacy and the Arab Digital Activism," September 17, 2010, nawaat.org/portail/2010/09/17/the -internet-freedom-fallacy-and-the-arab-digital-activism.

195 **While the Bahraini government was arresting bloggers and suppressing dissent, the United States was planning to sell $70 million in arms to Bahrain:** See Ivan Sigal, "Going Local," *Index on Censorship* 40, no. 1 (2011): 93–99.

195 **When Clinton visited Cairo a month after the revolution, Egypt's January 25 Revolution Youth Coalition refused to meet with her:** Kirit Radia and Alex Marquardt, "Young Leaders of Egypt's Revolt Snub Clinton in Cairo," ABC News Political Punch, March 15, 2011, http://blogs.abcnews.com /politicalpunch/2011/03/young-leaders-of-egypts-revolt-snub-clinton -in-cairo.html.

195 **"International Strategy for Cyberspace":** www.whitehouse.gov/sites/default /files/rss_viewer/internationalstrategy_cyberspace.pdf.

196   Swedish Foreign Minister Carl Bildt called for a "new transatlantic part-
      nership for protecting and promoting the freedoms of cyberspace": Carl
      Bildt, "Tear Down These Walls Against Internet Freedom," *Washington Post*,
      January 25, 2010, www.washingtonpost.com/wp-dyn/content/article/2010
      /01/24/AR2010012402297.html.

196   In July 2010 the French and Dutch foreign ministers convened an interna-
      tional conference on the Internet and freedom of expression: "Ministers to
      Meet in the Netherlands to Champion Internet Freedom," Permanent
      Mission of the Kingdom of the Netherlands to the United Nations, nether
      landsmission.org/article.asp?articleref=AR00000996EN.

196   President Nicolas Sarkozy declared in a speech at the Vatican: "Regulate
      Internet, Says Sarkozy," The Connexion, October 2010, http://connexion
      france.com/Sarkozy-control-internet-immorality-pope-12150-view-article
      .html (accessed June 27, 2011).

197   La Quadrature du Net published a leaked letter from Sarkozy to French
      Foreign Minister Bernard Kouchner: "Sarkozy Exports Repressive Inter-
      net," La Quadrature du Net, October 21, 2010, www.laquadrature.net/en
      /sarkozy-exports-repressive-internet.

198   Council of Europe published two documents: Council of Europe, "Internet
      Freedom Conference—From Principles to Global Treaty Law?" www.coe.int
      /t/dghl/standardsetting/media-dataprotection/conf-internet-freedom.

199   UN Special Rapporteur on Freedom of Expression Frank La Rue delivered
      a report to the UN Human Rights Council: www2.ohchr.org/english/bodies
      /hrcouncil/docs/17session/a.hrc.17.27_en.pdf.

200   Barlow later declared from the stage: Videos of the e-G8 forum are at
      www.eg8forum.com/en.

201   June 2011, UNESCO published a report titled *Freedom of Connection—
      Freedom of Expression: The Changing Legal and Regulatory Ecology Shaping
      the Internet*: http://unesdoc.unesco.org/images/0019/001915/191594e.pdf.

201   the Council of Europe's commissioner for human rights, Thomas Ham-
      marberg, accused UNESCO of dodging responsibility: Owen Bowcott, "In-
      ternet Freedom 'Is a Matter for UN,'" *The Guardian*, June 17, 2011, www
      .guardian.co.uk/law/butterworth-and-bowcott-on-law/2011/jun/17/internet
      -freedom-matter-un (accessed June 27, 2011).

CHAPTER 13: GLOBAL INTERNET GOVERNANCE

204   the Tunisian government tried to cancel our event: For an account of the
      events of that day, see Rebecca MacKinnon, "WSIS: Defending Free Speech
      in Tunis," RConversation blog, November 17, 2005, http://rconversation
      .blogs.com/rconversation/2005/11/wsis_defending_.html.

205   reasons cited by the Chinese government in its 2005 bid: For an account of
      the politics leading up to the WSIS summit, see Laura DeNardis, *Protocol
      Politics: The Globalization of Internet Governance* (Cambridge, MA: MIT
      Press, 2009). For the early history of ICANN, see Milton Mueller, *Ruling*

the Root: Internet Governance and the Taming of Cyberspace (Cambridge, MA: MIT Press, 2004).

207 **IGF, if organized and managed well, has the potential to serve as a global "coral reef":** Milton Mueller, *Networks and States: The Global Politics of Internet Governance* (Cambridge, MA: MIT Press, 2010), 49–50, 125.

209 **warned me not to mention any UN member countries in my remarks:** A transcript of the session can be found at www.intgovforum.org/cms/2009 /sharm_el_Sheikh/Programme.MainSessions.html#Emerging. For my own account of what happened, see Rebecca MacKinnon, "Muzzled by the United Nations," RConversation blog, November 18, 2009, http://rconversation .blogs.com/rconversation/2009/11/muzzled-by-the-united-nations.html.

209 **the OpenNet Initiative had been confronted by UN security:** For a detailed account of the incident, see "FAQ: What Happened at the Internet Governance Forum?" OpenNet Initiative, http://opennet.net/faq-what-happened -internet-governance-forum.

209 **ICANN CEO Rod Beckstrom gave an impassioned speech:** For the full text of his speech, see http://icann.org/en/presentations/beckstrom-speech-united -nations-14dec10-en.pdf.

210 **In June 2009, I attended one of ICANN's public meetings:** Transcripts of all open proceedings are posted online. For that particular meeting, see http://syd.icann.org/syd/transcripts. For information about ICANN's schedule, more general information about its work, and how to participate remotely online, see www.icann.org.

214 **new generic top-level domains:** See www.icann.org/en/topics/new-gtld -program.htm.

216 **effort to force ICANN to delay the launch of the new gTLD program:** See Milton Mueller, "Competition Policy Letters to ICANN Part of a US-EC 'Plot,'" Internet Governance Project blog, June 19, 2011, http://blog.internet governance.org/blog/_archives/2011/6/19/4841358.html; and "Why the Board Must Move Ahead with the New TLD Program Despite the GAC's Objections," Internet Governance Project blog, June 19, 2011, http://blog .internetgovernance.org/blog/_archives/2011/6/19/4841838.html.

216 **"nation states are just actors at the table, not predominant":** For video of the interview with Eliot Noss, see Brenden Kuerbis, "Interview: The Real Meaning of the ICANN's New gTLD Vote," Internet Governance Project blog, July 8, 2011, http://blog.internetgovernance.org/blog/_archives/2011/7/8 /4854455.html.

217 **ICANN's Non-Commercial Stakeholder Group (NCSG):** See http:// gnso.icann.org/non-commercial; and http://ncdnhc.org.

218 **"Welcome to 'bottom-up' policy making at ICANN":** Robin Gross, "Is ICANN Accountable to the Global Public Interest?" IPJustice, July 13, 2009, http://ipjustice.org/ICANN/NCSG/ba-NCUC-ICANN-Injustices.html.

218 **"did this simply to appease the commercial user groups":** Milton Mueller, "Under Pressure from Trademark Interests, ICANN Undoes the GNSO Re-

forms," Internet Governance Project blog, July 13, 2009, http://blog.internet governance.org/blog/_archives/2009/7/13/4253458.html.

219   Milton Mueller points out that multi-stakeholder institutions like ICANN have their own form of international, multi-stakeholder "pluralist politics": Mueller, *Networks and States*, p. 266.

CHAPTER 14: BUILDING A NETIZEN-CENTRIC INTERNET

221   **"We are all Samir Feriani":** Afef Abrougui, "Tunisia: Protest to Free a Government Critic," Global Voices Online, June 13, 2011, http://globalvoices online.org/2011/06/13/tunisia-a-protest-to-free-a-government-critic.

223   **Rosental Alves . . . likes to compare the pre-Internet age to a desert:** Rebecca MacKinnon, "Global Voices: Building Sustainable Civilization in an Information Rainforest," RConversation blog, May 9, 2010, http://rconversation.blogs .com/rconversation/2010/05/global-voices-building-sustainable-civilization -in-an-information-rainforest.html. Also see www.knightfoundation.org /staff/rosental-c-alves.

225   **Wikipedia:** See Andrew Lih, *The Wikipedia Revolution: How a Bunch of Nobodies Created the World's Greatest Encyclopedia* (New York: Hyperion, 2009).

226   **Global Voices Advocacy:** http://advocacy.globalvoicesonline.org.

226   **Tactical Technology Collective:** www.tacticaltech.org.

226   **Mobile Active:** www.mobileactive.org.

227   **Speak to Tweet:** http://twitter.com/#!/speak2tweet.

227   **Telecomix:** www.telecomix.org.

228   **New America Foundation's Open Technology Initiative:** http://oti .newamerica.net.

228   **Commotion Wireless:** http://oti.newamerica.net/commotion_wireless_0; and http://tech.chambana.net/projects/commotion. Regarding State Department funding, see James Glanz and John Markoff, "US Underwrites Internet Detour Around Censors," *New York Times*, June 12, 2011, www .nytimes.com/2011/06/12/world/12internet.html.

228   **Serval:** www.servalproject.org.

229   **"anonymizer" tool called Tor:** https://www.torproject.org. Disclosure: I served on its board of directors for one year in 2007.

230   **Diaspora:** See "Taking a Look at Social Network Diaspora," NY Convergence, March 14, 2011, http://nyconvergence.com/2011/03/taking-a-look -at-social-network-diaspora.html.

230   **Crabgrass:** http://crabgrass.riseuplabs.org.

230   **StatusNet:** http://status.net.

230   **FreedomBox:** https://freedomboxfoundation.org. Also see Jim Dwyer, "Decentralizing the Internet So Big Brother Can't Find You," *New York Times*, February 15, 2011, www.nytimes.com/2011/02/16/nyregion/16about.html; and "Freedom in the Cloud: Software Freedom, Privacy, and Security for Web 2.0 and Cloud Computing—A Speech Given by Eben Moglen at a

Meeting of the Internet Society's New York Branch on Feb. 5, 2010," Software Freedom Law Center, www.softwarefreedom.org/events/2010/isoc-ny/FreedomInTheCloud-transcript.html.

232 **Chaos Computer Club:** www.ccc.de/en; for a colorful description of the CCC's characters and culture, see Becky Hogge, *Barefoot into Cyberspace: Adventures in Search of Techno-Utopia* (London: Rebecca Hogge, 2011).

232 **Chaos Communication Camp:** http://events.ccc.de/camp/2011.

232 **yearly winter conferences:** See http://events.ccc.de/congress/2010/wiki/Welcome.

232 **"A Declaration of the Independence of Cyberspace":** https://projects.eff.org/~barlow/Declaration-Final.html.

233 **Douglas Rushkoff called on the netizens of the world to unite:** Douglas Rushkoff, "The Next Net," Shareable.net, January 3, 2011, http://shareable.net/blog/the-next-net. Also see his most recent book, *Program or Be Programmed: Ten Commands for a Digital Age* (New York: OR Books, 2010).

233 **"The invention of a tool doesn't create change":** Clay Shirky, *Here Comes Everybody: The Power of Organizing Without Organizations* (New York: Penguin Press, 2008), 105.

233 **"cute-cat theory of digital activism":** Ethan Zuckerman, "The Cute Cat Theory Talk at ETech," My Heart's in Accra blog, March 8, 2008, www.ethanzuckerman.com/blog/2008/03/08/the-cute-cat-theory-talk-at-etech.

234 **in 2007 WITNESS launched its own Video Hub:** http://hub.witness.org; Yvette Alberdingk Thijm, "Update on the Hub and WITNESS' New Online Strategy," August 18, 2010, http://blog.witness.org/2010/08/update-on-the-hub-and-witness-new-online-strategy; Ethan Zuckerman, "Public Spaces, Private Infrastructure—Open Video Conference," My Heart's in Accra blog, October 1, 2010, www.ethanzuckerman.com/blog/2010/10/01/public-spaces-private-infrastructure-open-video-conference.

234 **"Protecting Yourself, Your Subjects and Your Human Rights Videos on YouTube":** http://youtube-global.blogspot.com/2010/06/protecting-yourself-your-subjects-and.html.

234 **2010 Global Voices Citizen Media Summit:** Sami Ben Gharbia, "GV Summit 2010 Videos: A Discussion of Content Moderation," Global Voices Advocacy, May 7, 2010, http://advocacy.globalvoicesonline.org/2010/05/07/gv-summit-2010-videos-a-discussion-of-content-moderation; and Rebecca MacKinnon, "Human Rights Implications of Content Moderation and Account Suspension by Companies," RConversation blog, May 14, 2010, http://rconversation.blogs.com/rconversation/2010/05/human-rights-implications.html;

235 **"Digital Maoism":** Jaron Lanier, "Digital Maoism: The Hazards of the New Online Collectivism," Edge: The Third Culture, May 30, 2006, www.edge.org/3rd_culture/lanier06/lanier06_index.html. Also see Jaron Lanier, *You Are Not A Gadget: A Manifesto* (New York: Random House, 2010).

238 **Students for Free Culture:** http://freeculture.org.

238 In 2009 Sweden's Pirate Party won two seats in the European Parliament: Tom Sullivan, "Sweden's Pirate Party Sets Sail for Europe," *The Christian Science Monitor*, June 8, 2009, www.csmonitor.com/World/Europe/2009/0608/p06s08-woeu.html (accessed August 15, 2011).

238 green parties have taken up Internet freedom: German Green Party politician Malte Spitz, for example, has taken up the fight against surveillance and censorship as a signature issue. See Noam Cohen, "It's Tracking Your Every Move and You May Not Even Know," *New York Times*, March 26, 2011, www.nytimes.com/2011/03/26/business/media/26privacy.html (accessed August 15, 2011).

240 Charter of Human Rights and Principles for the Internet: See http://internetrightsandprinciples.org.

242 OECD ... "Communiqué on Principles for Internet Policymaking": Downloadable at www.oecd.org/dataoecd/40/21/48289796.pdf. Meeting description and other materials can be found at: www.oecd.org/site/0,3407,en_2157 1361_47081080_1_1_1_1_1,00.html.

242 civil society . . . could not endorse: http://csisac.org/CSISAC_Statement _on_OECD_Communique_06292011_FINAL_COMMENTS.pdf; press release: http://csisac.org/CSISAC_PR_06292011.pdf.

243 members of the Obama administration . . . praised the principles: Karen Kornbluh and Daniel J. Weitzner, "Foreign Policy on the Internet," July 14, 2011, www.washingtonpost.com/opinions/foreign-policy-of-the-internet/2011 /07/08/gIQAjqFyEI_story.html (accessed August 15, 2011).

244 Google took a step in this direction by launching . . . the Transparency Report: www.google.com/transparencyreport.

245 Measurement Lab, or M-Lab: www.measurementlab.net.

246 "Individuals need to be at the centers of their own digital lives": Rick Levine, Christopher Locke, Doc Searls, and David Weinberger, *The Cluetrain Manifesto: 10th Anniversary Edition* (New York: Basic Books, 2009), 16. Also see http://blogs.law.harvard.edu/vrm.

247 a report by the World Economic Forum: World Economic Forum, *Personal Data: The Emergence of a New Asset Class*, January 2011, www.weforum.org /reports/personal-data-emergence-new-asset-class.

247 *Democratizing Innovation*: Eric von Hippel, *Democratizing Innovation* (Cambridge, MA: MIT Press, 2005).

249 I visited Ai Weiwei at his home: Rebecca MacKinnon, "Ai Weiwei: On Taking Individual Responsibility," RConversation blog, January 22, 2009, http://rconversation.blogs.com/rconversation/2009/01/conversation-with -ai-weiwei.html.

AFTERWORD TO THE PAPERBACK EDITION

256 range of victories against government: Katie Stallard, "Snoopers' Charter: Nick Clegg Urges Re-Think," Sky News, December 11, 2012, accessed December 21, 2012, at http://news.sky.com/story/1023557/snoopers-charter

-nick-clegg-urges-re-think; Hasan Lakkis, "Aoun: Providing SMS data unconstitutional," *Daily Star*, December 5, 2012, accessed December 21, 2012, at http://www.dailystar.com.lb/News/Politics/2012/Dec-05/197253-aoun -providing-sms-data-unconstitutional.ashx#axzz2EGkelc1i; "Philippines court suspends cybercrime law," Radio Australia, October 9, 2012, accessed December 21, 2012, at http://www.radioaustralia.net.au/international /2012-10-09/philippines-court-suspends-cybercrime-law/1027900; Rebecca MacKinnon, "Fighting the Great Firewall of Pakistan," *Foreign Policy*, April 10, 2012, accessed December 21, 2012, at http://www.foreignpolicy.com /articles/2012/04/10/fighting_the_great_firewall_of_pakistan.

257 **For every half-year reporting cycle in between:** See http://www.google.com /transparencyreport/userdatarequests/ (accessed December 21, 2012).

259 **proposed a Wireless Surveillance Act:** "Markey: Law Enforcement Collecting Information on Millions of Americans from Mobile Phone Carriers," Website of Congressman Ed Markey accessed December 21, 2012, at http://markey.house.gov/press-release/markey-law-enforcement-collecting -information-millions-americans-mobile-phone-carriers; and Julie Ershadi, "Markey confronts law enforcement, DOJ over growing cellphone record requests," *Hillicon Valley*, August 10, 2012, accessed December 21, 2012, at http://thehill.com/blogs/hillicon-valley/technology/243127-rep-markey -confronts-law-enforcement-doj-over-growing-cellphone-record-requests.

259 **In November 2012 the *New York Times* reported that New York police:** Joseph Goldstein, "City Is Amassing Trove of Cellphone Logs," *New York Times*, November 26, 2012, accessed December 21, 2012, at http:// www.nytimes.com/2012/11/27/nyregion/new-york-city-police-amassing-a -trove-of-cellphone-logs.html.

260 **the Swedish investigative television program *Uppdrag Granskning*:** Ryan Gallagher, "Your Eurovision Song Contest Vote May Be Monitored: Mass Surveillance in Former Soviet Republics," Slate.com, April 30, 2012, accessed December 21, 2012, at http://www.slate.com/blogs/future_tense/2012/04/30 /black_box_surveillance_of_phones_email_in_former_soviet_republics_.html.

260 **among the countries that have embraced:** Pearce, K. E., and Kendzior, S. (2012), "Networked Authoritarianism and Social Media in Azerbaijan." *Journal of Communication*, 62: 283–298.

261 **"mobile phone system in Belarus is better for the opposition than for the regime":** "TeliaSonera 'profits by helping dictators spy,'" *The Local: Sweden's News in English*, April 18, 2012, accessed December 21, 2012, at http:// www.thelocal.se/40334/20120418/#.UNRwcLZEDqo.

262 **Throughout 2011 and 2012 GNI worked to build:** See https://global networkinitiative.org/news.

263 **2012 Pew Research Center survey:** "Arab Publics Most Likely to Express Political Views Online; Social Networking Popular Across Globe," Pew Research Center, December 12, 2012, accessed December 21, 2012, at http://www.pewglobal.org/2012/12/12/social-networking-popular-across -globe/.

264 **International Principles on Communications Surveillance and Human Rights:** Emma Draper, "PI is pleased to announce a public consultation on the International Principles on Communications Surveillance and Human Rights," Privacy International, December 11, 2012, accessed December 21, 2012, at https://www.privacyinternational.org/blog/pi-is-pleased-to-announce -a-public-consultation-on-the-international-principles-on.

266 **Activists seeking to force Facebook to abide by European privacy laws:** Somini Sengupta, "Should Personal Data Be Personal?" *New York Times*, February 4, 2012, accessed December 21, 2012, at http://www.nytimes.com /2012/02/05/sunday-review/europe-moves-to-protect-online-privacy.html.

266 **Schleswig-Holstein ruled that Facebook's real-name policy:** Natasha Lomas, "Facebook Users Must Be Allowed to Use Pseudonyms, Says German Privacy Regulator; Real-Name Policy 'Erodes Online Freedoms,'" December 18, 2012, accessed December 21, 2012, at http://techcrunch.com/2012 /12/18/facebook-users-must-be-allowed-to-use-pseudonyms-says -german-privacy-regulator-real-name-policy-erodes-online-freedoms/.

266 **Internet users' rights because the market would punish them:** "Technology giants at war: Another game of thrones," *The Economist*, December 1, 2012, accessed December 21, 2012, at http://www.economist.com/news/21567361 -google-apple-facebook-and-amazon-are-each-others-throats-all-sorts -ways-another-game.

INDEX